The Superworld I

THE SUBNUCLEAR SERIES

Series Editor: ANTONINO ZICHICHI, *European Physical Society, Geneva, Switzerland*

1. 1963 STRONG, ELECTROMAGNETIC, AND WEAK INTERACTIONS
2. 1964 SYMMETRIES IN ELEMENTARY PARTICLE PHYSICS
3. 1965 RECENT DEVELOPMENTS IN PARTICLE SYMMETRIES
4. 1966 STRONG AND WEAK INTERACTIONS
5. 1967 HADRONS AND THEIR INTERACTIONS
6. 1968 THEORY AND PHENOMENOLOGY IN PARTICLE PHYSICS
7. 1969 SUBNUCLEAR PHENOMENA
8. 1970 ELEMENTARY PROCESSES AT HIGH ENERGY
9. 1971 PROPERTIES OF THE FUNDAMENTAL INTERACTIONS
10. 1972 HIGHLIGHTS IN PARTICLE PHYSICS
11. 1973 LAWS OF HADRONIC MATTER
12. 1974 LEPTON AND HADRON STRUCTURE
13. 1975 NEW PHENOMENA IN SUBNUCLEAR PHYSICS
14. 1976 UNDERSTANDING THE FUNDAMENTAL CONSTITUENTS OF MATTER
15. 1977 THE WHYS OF SUBNUCLEAR PHYSICS
16. 1978 THE NEW ASPECTS OF SUBNUCLEAR PHYSICS
17. 1979 POINTLIKE STRUCTURES INSIDE AND OUTSIDE HADRONS
18. 1980 THE HIGH-ENERGY LIMIT
19. 1981 THE UNITY OF THE FUNDAMENTAL INTERACTIONS
20. 1982 GAUGE INTERACTIONS: Theory and Experiment
21. 1983 HOW FAR ARE WE FROM THE GAUGE FORCES
22. 1984 QUARKS, LEPTONS, AND THEIR CONSTITUENTS
23. 1985 OLD AND NEW FORCES OF NATURE
24. 1986 THE SUPERWORLD I

Volume 1 was published by W. A. Benjamin, Inc., New York; 2–8 and 11–12 by Academic Press, New York and London; 9–10, by Editrice Compositori, Bologna; 13-24 by Plenum Press, New York and London.

The Superworld I

Edited by
Antonino Zichichi
European Physical Society
Geneva, Switzerland

PLENUM PRESS • NEW YORK AND LONDON

Library of Congress Cataloging in Publication Data

International School of Subnuclear Physics (24th: 1986: Erice, Italy)
The superworld I / edited by Antonino Zichichi.
p. cm.—(The Subnuclear series; 24)
"Proceedings of the Twenty-fourth course of the International School of Subnuclear Physics on the superworld I, held August 7-15, 1986, in Erice, Sicily, Italy"—T.p. verso.
Includes bibliographical references.
ISBN-13: 978-1-4684-1320-5 e-ISBN-13: 978-1-4684-1318-2
DOI: 10.1007/978-1-4684-1318-2
1. Superstring theories—Congresses. 2. Supergravity—Congresses. I. Zichichi, Antonino. II. Title. III. Title: Superworld one. IV. Title: Superworld 1. V. Series: Subnuclear series; v. 24.
QC794.6.S85I57 1986 89-35914
539.7′2—dc20 CIP

Proceedings of the Twenty-Fourth Course of the International School of Subnuclear Physics on The Superworld I, held August 7-15, 1986, in Erice, Sicily, Italy

Softcover reprint of the hardcover 1st edition 1990
A Division of Plenum Publishing Corporation
233 Spring Street, New York, N.Y. 10013

PREFACE

During August 1986, a group of 96 physicists from 61 laboratories in 17 countries met in Erice for the 24th Course of the International School of Subnuclear Physics. The countries represented were: Austria, Bulgaria, Belgium, Denmark, France, the Federal Republic of Germany, Hungary, Israel, Italy, Poland, Portugal, South Africa, Spain, Sweden, Switzerland, United Kingdom, United States of America. The School was sponsored by the European Physical Society (EPS), the Italian Ministry of Education (MPI), the Sicilian Regional Government (ERS), and the Weizmann Institute of Science.

I hope the reader will enjoy this book as much as the students enjoyed attending the lectures and the discussion Sessions, which are the most attractive features of the School. Thanks to the work of the Scientific Secretaries, the discussions have been reproduced as faithfully as possible. At various stages of my work I have enjoyed the collaboration of my friends whose contributions have been extremely important for the School and are highly appreciated. I thank them most warmly. A final acknowledgement to all those who, in Erice, Bologna, Rome and Geneva, helped me on so many occasions and to whom I feel very much indebted.

Antonino Zichichi
November 1986
Geneva

CONTENTS

TWO-DIMENSIONAL SUPERSPACES*

P. van Nieuwenhuizen**

Instituut-Lorentz voor Theoretische Natuurkunde
University of Leiden, Leiden, The Netherlands

PREFACE

The following lectures deal with 2-dimensional superspaces. The constraints are derived and solved, the superconformal gauge and Weyl invariance are discussed and the Virasoro constraints are obtained.

Also discussed at the school but not contained in these proceedings are applications of these ideas to the measure of the path-integral in superspace. One can derive the critical dimension of strings following Fujikawa's method, provided one takes into account the constraints on the geometry in superspace.

A comprehensive discussion of Fujikawa's method in general can be found in the following two references:

K. Fujikawa, U. Lindstrom, N.K. Nielsen, M. Rocek and P. van Nieuwenhuizen Stony Brook preprint spring, 1987.
P. van Nieuwenhuizen. Lectures at the First Chilean Winter School, December, 1985.

For a derivation of the critical dimension in superspace using path integral techniques, see ref. (4). For critical dimensions of strings on group manifolds using path-integral measures, see:

A. Eastaugh, L. Mezincescu, E. Sezgin, and P. van Nieuwenhuizen , Phys. Rev. Lett. 57, 29 (1986).
A Ceresole, A. Lerda, P. Pizzochero, and P. van Nieuwenhuizen, Phys. Lett. B. to be published.

---

* Teyler's lectures at Leyden University, June, 1986.

**On leave from the Institute for Theoretical Physics at the State University of New York at Stony Brook, Stony Brook, N.Y. 11794, U.S.A.

1. Two-dimensional superspaces: introduction

Two-dimensional superspaces exhibit similar features as four-dimensional superspaces, but two-dimensional superspaces are simpler and of paedagogical value. Moreover, with the advent of string theories, which are based on two-dimensional supergravities, two-dimensional superspaces may become of physical interest as well.

The simplest (N=1) two-dimensional superspaces have two bosonic coordinates x^0 and x^1, and two fermionic (anticommuting) coordinates θ^1 and θ^2. In addition there are superspaces with four θ's (the N=2 models) and with eight θ's (the N=4 models), but we will not discuss these more complicated models. The fermionic coordinates are Grassmann variables,

$$\{\theta^\alpha, \theta^\beta\} = \theta^\alpha\theta^\beta + \theta^\beta\theta^\alpha = 0 . \tag{1}$$

The coordinate x^0 is a time coordinate, while x^1 is a space coordinate; often one denotes them by τ and σ, respectively. For fermionic coordinates such a distinction is not present because $\theta^1\theta^1 = \theta^2\theta^2 = 0$.

The two geometrical fields defined on the x-θ space are the supervielbein

$$E_\Lambda{}^M(x,\theta); \quad \Lambda = \{\mu,\alpha\} , \quad M = \{m,a\} \tag{2}$$

and the superconnection for local tangent transformations

$$\Omega_{\Lambda M}{}^N(x,\theta); \quad \Lambda, M \text{ and } N \text{ as in (2).} \tag{3}$$

The two local symmetry transformations are

(i) local supercoordinate transformations with superparameter

$$E^\Lambda(x,\theta) = \{E^\mu(x,\theta), \ E^\alpha(x,\theta)\} \tag{4}$$

(ii) local tangent transformations with superparameter

$$L_M{}^N(x,\theta) \tag{5}$$

The supervielbein and superconnection transform per definition as follows

$$\delta E_\Lambda{}^M = E^\Pi(\partial_\Pi E_\Lambda{}^M) + (\partial_\Lambda E^\Pi)E_\Pi{}^M + E_\Lambda{}^N L_N{}^M \tag{6}$$

$$\delta\Omega_{\Lambda M}{}^N = E^\Pi(\partial_\Pi\Omega_{\Lambda M}{}^N) + (\partial_\Lambda E^\Pi)\Omega_{\Pi M}{}^N - \mathcal{D}_\Lambda L_M{}^N , \tag{7}$$

where the last term is the covariant derivative of the tangent parameter

$$\mathcal{D}_\Lambda L_M{}^N = \partial_\Lambda L_M{}^N + \Omega_{\Lambda M}{}^P L_P{}^N - L_M{}^P \Omega_{\Lambda P}{}^N (-)^{\Lambda(M+P)} . \tag{8}$$

The symbol $(-)^{\Lambda}$ is +1 if Λ refers to a bosonic index, and -1 if it refers to a fermionic index. It would have been more precise to write $(-)^{\sigma(\Lambda)}$ with $\sigma(\Lambda) = +1$ or -1, but to avoid cumbersome notation, we will omit these σ's.

We will always contract superindices in north-westerley fashion: from the left-upper to the right-lower position. The origin of the sign factor in (8) is now clear: after bringing Λ to the far left, the the P contraction satisfies the north-west rule, but in bringing Λ to the far left, it passes the indices M and P, and picks up the sign factor in (8).

Up to this point, we have followed the ideas of general relativity: we defined an arbitrary supervielbein and superconnection on a given manifold. We could now go on and try to find a gauge action, but this would not be possible since, as we shall see, in two-dimensional superspaces the gauge action is a total derivative, just as the Einstein-Hilbert action $\sqrt{g}R$ in two-dimensional spacetime in a total derivative (namely the topological Euler invariant). Worse, the supervielbein cannot be taken to be an arbitrary function of x,θ, but must satisfy certain constraints, as we shall discuss. The nature of these constraints can be understood if one considers the two-dimensional superspaces as coset manifolds, and therefore we shall first give a discussion of coset manifolds. We begin by defining the relevant superalgebra which defines the particular coset manifold we are interested in.

2. The super Poincaré algebra

The two-dimensional super Poincaré algebra contains in addition to the two translation generators P_0 and P_1 and the one Lorentz generator J, two anticommuting generators Q^a (a=1,2). The (anti)commutation relations are given by $[X_A, X_B\} = f_{AB}{}^C X_C$, where the symbol $[,\}$ denotes a commutator if A and/or B are bosonic quantities, and an anticommutator if both A and B are fermionic. In detail

$$
\begin{aligned}
&[P_m, P_n] = 0; \qquad [P_0, J_{01}] = -P_1 \qquad (J_{01} \equiv J) \\
&\qquad\qquad\qquad\quad [P_1, J_{01}] = -P_0 \\
&[Q^a, P_m] = 0; \qquad [Q^a, J_{01}] = \tfrac{1}{2}(\gamma_0\gamma_1)^a{}_b\, Q^b \\
&\{Q^a, Q^b\} = 2i(\gamma^m)^{ab}\, P_m \,.
\end{aligned}
\tag{9}
$$

The Dirac matrices γ^0 and γ^1 are 2x2 Pauli matrices, and we choose the following representation

$$(\gamma^0)^a{}_b = \begin{pmatrix} 0 & -1 \\ +1 & 0 \end{pmatrix}, \quad (\gamma^1)^a{}_b = \begin{pmatrix} 0 & 1 \\ 1 & 0 \end{pmatrix}. \tag{10}$$

It follows that $\gamma_0\gamma_1 = +\tau_3 = \begin{pmatrix} +1 & 0 \\ 0 & -1 \end{pmatrix}$. Hence, $[Q^1,J] = \frac{1}{2}Q^1$, and $[Q^2,J] = -\frac{1}{2}Q^2$. The Dirac matrices satisfy the Clifford algebra

$$\{\gamma^m,\gamma^n\} = 2\eta^{mn}; \quad \eta^{mn} = (-1,+1) \quad \text{for } m=0,1. \tag{11}$$

We shall raise and lower vector indices m,n by the Minkovski metric η_{mn}, and spinor indices a,b by the charge conjugation matrix C_{ab}. The latter satisfies the following relations

$$C^T = -C, \quad C\gamma^m C^{-1} = -(\gamma^m)^T. \tag{12}$$

It is clear that C_{ab} is proportional to ε_{ab}, and we shall take it equal to ε_{ab}. We shall again raise and lower indices in north-westerly fashion. For example

$$(\gamma^m)^{ab} = (\gamma^m)^a{}_c\,\varepsilon^{bc} \quad \text{and} \quad \gamma^m{}_{ab} = \varepsilon_{ca}(\gamma^m)^c{}_b, \tag{13}$$

where $\varepsilon_{12} = \varepsilon^{12} = +1$. The reader may verify that the Dirac matrices with both indices up or down are diagonal

$$\gamma^0_{ab} = -I; \quad \gamma_0^{ab} = +I;$$

$$\gamma^1{}_{ab} = -\tau_3 = \begin{pmatrix} -1 & 0 \\ 0 & +1 \end{pmatrix}; \quad \gamma_1^{ab} = +\tau_3 = \begin{pmatrix} 1 & 0 \\ 0 & -1 \end{pmatrix}. \tag{14}$$

With these definitions one has

$$[P_m,J] = -\varepsilon_m{}^n P_n, \quad \varepsilon_m{}^n = \begin{pmatrix} 0 & +1 \\ +1 & 0 \end{pmatrix}, \tag{15}$$

where $\varepsilon_m{}^n = \varepsilon_{ms}\eta^{sn}$ for $m,n=0,1$, and $\varepsilon_{01} = +1$. Note that ε_{mn} acts on vectors, but ε_{ab} on spinors.

The generators Q^a are Majorana spinors, which means that their Majorana conjugate Q^TC is equal to their Dirac conjugate $Q^\dagger\gamma_0$. Since $(\gamma_0)^a{}_b = \varepsilon_{ab} = C_{ab}$, it follows that the Q^a are hermitian. The bosonic generators we will take antihermitian, in order that the structure constants in the purely bosonic sector (the Poincaré algebra) are real. It is then inevitable that in the fermi-fermi anticommutators the structure constants are purely imaginary. This explains the factor i in the $\{Q,Q\}\sim P$ anti-commutator.

From the definition of $\bar{Q}_a = Q^b C_{ba}$, one may verify the following results for $\bar{Q}_a$

$$\{\bar{Q}_a,\bar{Q}_b\} = 2i\gamma^m{}_{ab}P_m \; ; \quad [\bar{Q}_a,P_m] = 0$$

$$[\bar{Q}_a,J] = -\tfrac{1}{2}(\tau_3)_a{}^b\bar{Q}_b, \quad (\tau_3)_a{}^b = \begin{pmatrix} 1 & 0 \\ 0 & -1 \end{pmatrix} \tag{16}$$

Note that $(\tau_3)^a{}_b = (\tau_3)^b{}_a$.

If one introduces so-called light-cone coordinates for vectors,

$$P_{++} \equiv -(P_0+P_1), \quad P_{--} \equiv -P_0+P_1 \tag{17}$$

then one may note that

$$[P_{++},J] = -P_{++}; \quad [P_{--},J] = P_{--} \; . \tag{18}$$

Since $Q^1 = \tfrac{1}{2}\big((1+\tau_3)Q\big)^1$ and $Q^2 = \tfrac{1}{2}\big((1-\tau_3)Q\big)^2$ are one-dimensional representations of the Lorentz group with eigenvalues $+\tfrac{1}{2}$ and $-\tfrac{1}{2}$, respectively, we denote Q^1 by Q^+ and Q^2 by Q^-. It follows that $Q^+ = \bar{Q}_- = \bar{Q}_2$ and $Q^- = -\bar{Q}_+ = -\bar{Q}_1$. because $C_{12} = \varepsilon_{+-}$. Hence

$$[\bar{Q}_+,J] = -\tfrac{1}{2}\bar{Q}_+ \; ; \quad [\bar{Q}_-,J] = \tfrac{1}{2}\bar{Q}_- \; . \tag{19}$$

In fact, the transition to light-cone coordinates is nothing else but the van der Waerden formalism in two dimensions. Defining

$$v_{ab} \equiv v^m(\gamma_m)_{ab} = \begin{pmatrix} v^0-v^1 & 0 \\ 0 & v^0+v^1 \end{pmatrix} = \begin{pmatrix} -v_0-v_1 & 0 \\ 0 & -v_0+v_1 \end{pmatrix}$$

$$v^{ab} \equiv v^m(\gamma_m)^{ab} = \begin{pmatrix} v^0+v^1 & 0 \\ 0 & v^0-v^1 \end{pmatrix} \tag{20}$$

we see that $v^{++} = v_{--}$ and $v^{--} = v_{++}$. Moreover, $[v_{++},J] = [-(v_0+v_1),J] = -v_{++}$, in agreement with $[\bar{Q}_+,J] = -\tfrac{1}{2}\bar{Q}_+$. We shall often omit the bars on $\bar{Q}_a$; for example, $Q_+ = -Q^-$ and $Q_- = Q^+$.

The use of + and − indices is very convenient in two dimensions, because it reveals manifestly the Lorentz properties of tensors.

We will now verify that the superPoincaré algebra is consistent, i.e., the superJacobi identities are satisfied. In general the superJacobi identities are given by

$$[A,[B,C\}\} + [B,[C,A\}\} (-)^{A(B+C)} + [C,[A,B\}\} (-)^{C(A+B)} = 0 . \qquad (21)$$

Taking for A,B,C the generators P_m, Q^a and J, most identities are trivially satisfied. The only nontrivial identity is

$$\{Q^a,[Q^b,J]\} - \{Q^b,[J,Q^a]\} + [J,\{Q^a,Q^b\}] = 0 \qquad (22)$$

Substitution of the (anti)commutation relations yields

$$2i(\gamma^m)^{ac}\tfrac{1}{2}(\tau_3)^b{}_c - 2i(\gamma^m)^{bc}(-\tfrac{1}{2}\tau_3)^a{}_c + 2i(\gamma^n)^{ab}\ \varepsilon_n{}^m = 0 . \qquad (23)$$

Using that $\tau_3\gamma^m = -\gamma^{1-m}$ and $\gamma^n\varepsilon_n{}^m = \gamma^{1-m}$ for m=0,1, one sees that this identity is satisfied. Hence, the generators of the superPoincaré algebra form a closed algebraic system under commutators and anticommutators.

One could also study the super anti-de Sitter algebra, in which $[P_m,P_n] \sim \varepsilon_{mn}J$, or consider extra generators (central charges) which appear in $\{Q^{i,a},Q^{j,b}\} = \delta^{ij}(\gamma^m)^{ab}P_m + (C)^{ab}U^{ij} + (C\tau_3)^{ab}V^{ij}$. One could then study the most general form such extensions can take. We shall not do that here. For us, the important fact is that the superPoincaré algebra in (9) exists.

3. The geometry of rigid two-dimensional superspace

We consider as coset manifold the quotient of the superPoincaré algebra divided by the Lorentz algebra. Since the Lorentz algebra is a subalgebra of the superPoincaré algebra, this is a possible coset manifold. There are other coset manifolds one could consider, such as dividing by the subalgebra generated by P_m and $\bar{Q}_a$, but dividing by the Lorentz group will lead to fields which depend on x^μ and θ^α (superfields), and those are the fields we are interested in. One defines the inverse supervielbein and H-connection of <u>any</u> coset manifold G/H by

$$(\exp z^\Lambda K_\Lambda)(\exp dz^M K_M) =$$

$$\exp(z^\Lambda + dz^M \overset{\circ}{E}_M{}^\Lambda) K_\Lambda \exp(-dz^M E_M{}^\Lambda \overset{\circ}{\Omega}_\Lambda{}^i H_i) \ . \tag{24}$$

The superscript ° denotes that these are quantities of the rigid coset manifold: well-defined functions. The K_Λ are the coset generators, z^Λ the coset coordinates, and H_i equal to the Lorentz generator J, and

$$K_M = \{P_m, \bar{Q}_a\}; \quad z^\Lambda = \{x^\mu, \theta^\alpha\} \ . \tag{25}$$

Note that the products $z^\Lambda K_\Lambda$ and $dz^M K_M$ are always bosonic. Thus we can apply the standard Baker-Campbell-Hausdorff formula. Although there is no fundamental distinction between curved (super)indices Λ and flat (super)indices M on a coset manifold, a useful notational convention is to use flat indices for objects near the origin (z^Λ=0), and curved indices away from the origin. Hence, we write $z^\Lambda K_\Lambda$ and $dz^M K_M$. It follows from this convention that the supervielbein has one curved and one flat index: $\overset{\circ}{E}_M{}^\Lambda(x,\theta)$.

Using the general formula

$$\exp A \exp \varepsilon B = \exp\left(A+\varepsilon B + \varepsilon \sum_{n=1}^{\infty} \frac{1}{2n!} \underbrace{[A,\ldots,[A,B],\ldots,]}_{n\text{-times}} + \mathcal{O}(\varepsilon^2)\right) \tag{26}$$

we can compute the supervielbein $\overset{\circ}{E}_\Lambda{}^M(x,\theta)$ and H-connection $\overset{\circ}{\Omega}_\Lambda{}^i(x,\theta)$. Usually, one has contributions for all n, but in our case we will only get contributions for n=1, and everything can be evaluated in closed form.

Indeed, using

$$z^\Lambda K_\Lambda = x^\mu P_\mu + \theta^\alpha \bar{Q}_\alpha \ , \quad dz^M K_M = dx^m P_m + d\theta^a \bar{Q}_a$$

$$\tfrac{1}{2}[\theta^\alpha \bar{Q}_\alpha, d\theta^a \bar{Q}_a] = \tfrac{1}{2} d\theta^a \theta^\alpha (2i\gamma^\mu)_{\alpha a} P_\mu = d\theta^a (i\theta^\alpha \gamma^\mu{}_{\alpha a}) P_\mu \tag{27}$$

we find

$$\exp(x^\mu P_\mu + \theta^\alpha \bar{Q}_\alpha)\exp(dx^m P_m + d\theta^a \bar{Q}_a) =$$

$$\exp\left(x^\mu P_\mu + \theta^\alpha \bar{Q}_\alpha + dx^m\{\delta_m{}^\mu\}P_\mu + d\theta^a\{\delta_a{}^\alpha\}\bar{Q}_\alpha + d\theta^a\{i\theta^\alpha\gamma^\mu{}_{\alpha a}\}P_\mu\right) . \tag{28}$$

Hence, we find for the inverse supervielbein

$$\overset{\circ}{E}_M{}^\Lambda = \begin{pmatrix} \delta_m{}^\mu & 0 \\ i\theta^\alpha \gamma^\mu{}_{\alpha a} & \delta_a^\alpha \end{pmatrix} \tag{29}$$

while the H-connection vanishes

$$\overset{\circ}{\Omega}_\Lambda = 0 \tag{30}$$

since there is no term $\exp -dz^M \overset{\circ}{E}_M{}^{\Lambda} \overset{\circ}{\Omega}_\Lambda J$ in the product (28). This is due to the semi-direct sum into which the superPoincaré algebra splits.

The covariant derivatives on <u>any</u> coset manifold are defined by

$$\overset{\circ}{D}_M = \overset{\circ}{E}_M{}^{\Lambda}(\partial_\Lambda + \overset{\circ}{\Omega}_\Lambda{}^i H_i) \; . \tag{31}$$

When they act on a tensor in an H-representation R, one is supposed to take for H_i the H_i generators in that representation R. However, in our case $\overset{\circ}{\Omega}_\Lambda{}^i H_i = \overset{\circ}{\Omega}_\Lambda J = 0$, hence we find the following covariant derivatives

$$\overset{\circ}{D}_m = \partial_m, \quad \overset{\circ}{D}_a = \partial/\partial\theta^a + i\theta^c \gamma^\mu{}_{ca}\partial_\mu = \partial/\partial\theta^a - i(\bar{\theta}\not{\partial})_a \; . \tag{32}$$

These rigid covariant derivatives always satisfy the same algebra as the abstract generators

$$[\overset{\circ}{D}_M, \overset{\circ}{D}_N\} = f_{MN}{}^P \overset{\circ}{D}_P \; . \tag{33}$$

In our case we have indeed, as one may check explicitly,

$$[\overset{\circ}{D}_m, \overset{\circ}{D}_n] = 0 \; , \quad [\overset{\circ}{D}_m, \overset{\circ}{D}_a] = 0 \; , \quad \{\overset{\circ}{D}_a, \overset{\circ}{D}_b\} = 2i\gamma^m{}_{ab}\overset{\circ}{D}_m \; . \tag{34}$$

<u>Exercise</u>: Show that $\partial/\partial\theta^a$ is hermitian and $\partial/\partial x^\mu$ antihermitian (consider $\{\partial/\partial\theta^a, \theta^b\} = \delta_a{}^b$). Prove that $(D_+)^\dagger = D_+$ and that $D^2 \equiv D_a D^a$ is anti-hermitian. Derive the identity

$$D^2 D_b = -D_b D^2 = 2i\gamma^m{}_{bc} D^c \partial_m \; . \tag{35}$$

4. <u>The geometry of local two-dimensional superspace</u>

If one goes from rigid supersymmetry to local supersymmetry, the inverse supervielbein $E_M{}^\Lambda(x,\theta)$ and superconnection $\Omega_{\Lambda M}{}^N(x,\theta)$ are no longer well-defined functions of (x,θ). In general relativity, $e_m{}^\mu(x)$ and $\omega_{\mu m}{}^n(x)$ become even arbitrary functions of x *), but the situation is more interesting and less trivial in supergravity: the vielbein and connection must satisfy certain constraints. These constraints are imposed from the outside and do not follow from field equations; modulo these constraints, $E_M{}^\Lambda(x,\theta)$ and $\Omega_{\Lambda M}{}^N(x,\theta)$ are arbitrary functions. The constraints can be solved, and the independent superfields of which $E_M{}^\Lambda$ and $\Omega_{\Lambda M}{}^N$ are functions, are called prepotentials. The choice of constraints is a delicate matter, which we will

*) In the first-order formalism of general relativity, $\omega_{\mu m}{}^n(x)$ is arbitrary, but its algebraic field equation determines it as a function of $e_\mu{}^m(x)$, so-called Palatini formalism.

discuss. The local supervielbein $E_M{}^\Lambda$ and $\Omega_{\Lambda M}{}^N$ can vary freely (modulo the constraints) around their background values $\overset{\circ}{E}_M{}^\Lambda$ and $\overset{\circ}{\Omega}_{\Lambda M}{}^N$ of the coset manifold. Thus one must make sure that the constraints one imposes contain $\overset{\circ}{E}$ and $\overset{\circ}{\Omega}$ as solutions: the local geometrical objects oscillate around the geometry of the coset manifold.

Covariant derivatives of local supersymmetry are defined by

$$D_M = E_M{}^\Lambda(\partial_\Lambda + \Omega_\Lambda{}^i H_i) \ . \tag{36}$$

Thus, instead of a general local tangent connection $\Omega_{\Lambda M}{}^N(x,\theta)$, we use a local H-valued connection. <u>We specialize the tangent group to the Lorentz group</u>. This is not necessary, but in the spirit of the coset manifold G/H, where also the rigid connection was H-valued.

In our case, the covariant derivative D_M acts on tangent supervectors such as $E_N{}^\Pi$ as $D_M E_N{}^\Pi = E_M{}^\Lambda(\partial_\Lambda E_N{}^\Pi + \Omega_{\Lambda N}{}^P E_P{}^\Pi)$ where, as mentioned above, $\Omega_{\Lambda N}{}^P$ is Lorentz-valued. This means that $\Omega_{\Lambda N}{}^P = \Omega_\Lambda (J)_N{}^P$ where $(J)_N{}^P$ is the Lorentz generator in the coset-representation of the Lorentz group

$$(J)_N{}^P = \left(\begin{array}{c|c} \varepsilon_m{}^n & 0 \\ \hline 0 & \tfrac{1}{2}(\tau_3)_a{}^b \end{array}\right) = \left(\begin{array}{cc|cc} 0 & 1 & 0 & \\ 1 & 0 & & \\ \hline & 0 & \tfrac{1}{2} & 0 \\ & & 0 & -\tfrac{1}{2} \end{array}\right) \tag{37}$$

The local covariant derivatives define the supertorsion and supercurvatures as follows

$$[D_M, D_N\} = T_{MN}{}^P D_P + R_{MN}{}^i H_i \ . \tag{38}$$

In our case, $H_i = J$ and thus

$$[D_M, D_N\} = T_{MN}{}^P D_P + R_{MN} J \ . \tag{39}$$

We note that in general $\overset{\circ}{T}_{MN}{}^P = f_{MN}{}^P$ and $\overset{\circ}{R}_{MN}{}^i = f_{MN}{}^i$, but in our case $\overset{\circ}{R}_{MN} = 0$ and the only nonzero $\overset{\circ}{T}_{MN}{}^P$ is $f_{MN}{}^P = 2i\gamma_{ab}{}^m$.

The supertorsions and supercurvatures satisfy the Bianchi identities. They follow from the identity

$$[D_M,[D_N,D_P\}\} + (-)^{(N+P)M}[D_N,[D_P,D_M\}\} + (-)^{P(M+N)}[D_P,[D_M,D_N\}\} = 0 \ . \tag{40}$$

This is an <u>identity</u> since the 12 terms form 6 pairs and each pair cancels identically. Substituting (39), we find explicitly

$$D_M T_{NP}{}^R - T_{MN}{}^S T_{SP}{}^R - R_{MNP}{}^R + \text{supercyclic} = 0 \,. \tag{41}$$

$$D_M R_{NRS}{}^T - T_{MN}{}^P R_{PRS}{}^T + \text{supercyclic} = 0 \,, \tag{42}$$

where

$$R_{PRS}{}^T = R_{PR}{}^i (H_i)_S{}^T = R_{PR}(J)_S{}^T \,. \tag{43}$$

To derive this result we consider

$$[D_M, T_{NP}{}^R D_R + R_{NP} J\} = (D_M T_{NP}{}^R) D_R - (-)^{(N+P)M} T_{NP}{}^R [D_R, D_M\} + + (D_M R_{NP}) J + (-)^{M(N+P)} R_{NP} [D_M, J] \tag{44}$$

and using the supercyclicity the sign-factors can be removed. With

$$[D_M, J] = -J_M{}^N D_N \tag{45}$$

we arrive at (41) and (42).

<u>Exercise</u>: Show that R_{+-} is antihermitian and given by

$$E_+{}^\Lambda E_-{}^\Pi \big(\partial_\Lambda \Omega_\Pi (-)^{\Lambda(\Pi+1)} - (-)^\Lambda \partial_\Pi \Omega_\Lambda\big) =$$

$$= \partial_+ \Omega_- + \partial_- \Omega_+ - E_+{}^\Lambda (\partial_\Lambda E_-{}^\Pi) \Omega_\Pi - E_-{}^\Pi (\partial_\Pi E_+{}^\Lambda) \Omega_\Lambda$$

$$= \partial_+ \Omega_- + \partial_- \Omega_+ - T_{+-}{}^M \Omega_M - \Omega_+ \Omega_- \tag{46}$$

with $\Omega_\pm = E_\pm{}^\Lambda \Omega_\Lambda$ and $\partial_\pm = E_\pm{}^\Lambda \partial_\Lambda$.

5. <u>The constraints of local two-dimensional superspace</u>

In this section we shall explain why we take as constraints the following set of relations

$$T_{mn}{}^r = 0; \qquad T_{ab}{}^c = 0; \qquad T_{ab}{}^m = 2i\gamma^m{}_{ab} \,. \tag{47}$$

We shall give several unrelated arguments which all lead to these constraints, but there is no necessity to adopt these constraints.

The x-space formulation of supergravity contains a vielbein $e_\mu{}^m(x)$, a gravitino $\psi_\mu{}^a(x)$ and an auxiliary field $S(x)$, needed for a closed gauge algebra. Thus the number of field components minus the number of local x-space gauge invariances is 4, namely

$$
\begin{aligned}
&4e_\mu{}^m - 2 \text{ gen. coor.} - 1 \text{ local Lorentz} = 1 \quad \text{bose}\\
&4\psi_\mu{}^a - 2 \text{ local supersymmetry} = 2 \quad \text{fermi}\\
&1 \text{ auxiliary field } S(x) = 1 \quad \text{bose.}
\end{aligned}
\tag{48}
$$

These four x-space fields fill up one scalar superfield. (A scalar superfield $\phi(x,\theta)$ can be expanded as $A + \bar{\theta}\chi + F\bar{\theta}\theta$ and contains 2 bosonic and 2 fermionic field components). In superspace, the count is as follows:

$$
\begin{aligned}
&16E_\Lambda{}^M - 4 \text{ gen. supercoord.} - 1 \text{ local Lorentz} = 11 \text{ superfields}\\
&\Omega_{\Lambda M}{}^N = 4 \text{ superfields.}
\end{aligned}
\tag{49}
$$

We must thus impose <u>14 superfield constraints</u> in order to have in superspace the same number of x fields as in x-space. The set of relations in (47) consists indeed of 14 constraints:

$$
2T_{mn}{}^r, \quad 6T_{ab}{}^c, \quad 6T_{ab}{}^m . \tag{50}
$$

(recall that $T_{mn}{}^r$ is antisymmetric in mn, but $T_{ab}{}^c$ symmetric in ab).

The constraint $T_{mn}{}^r$ is the same as in general relativity the field equation for the spin connection. In superspace, $T_{mn}{}^r = 0$ will not be a field equation, but it determines $\Omega_{mn}{}^r$ in terms of $E_\Lambda{}^M(x,\theta)$, just as in general relativity. The constraint $T_{ab}{}^c$ not only determines $\Omega_{ab}{}^c$ but also constrains the supervielbein, as we shall see. In d=4 supersymmetric Yang-Mills theory, one constrains all fermionic components of the curvature to zero: $F_{ab} = 0$. The constraint $T_{ab}{}^c = 0$ is the analogue, but note that it is a constraint on the supertorsions. The last constraint, $T_{ab}{}^m = 2i\gamma^m{}_{ab}$ is in agreement with the coset manifold: $\overset{\circ}{T}_{ab}{}^m = 2i\gamma^m{}_{ab}$. Hence, <u>if</u> we constrain $T_{ab}{}^m$, it must be equal to $2i\gamma^m{}_{ab}$ and cannot vanish, because if $T_{ab}{}^m = 0$, the geometry of the coset manifold (rigid superspace) would not be included in the geometry of local superspace.

One can also use dimensional arguments to derive the above constraints. The geometric dimension of the x-space fields of supergravity is

$$[e_\mu{}^m] = 0; \quad [\psi_\mu{}^a] = ½; \quad [S] = 1. \tag{51}$$

In superspace, the geometric dimension of the supertorsions and supercurvatures is given by (see also (138))

$$\begin{aligned}
&[T_{ab}{}^m] = 0 \\
&[T_{am}{}^n] = [T_{ab}{}^c] = ½ \\
&[T_{ma}{}^b] = [T_{mn}{}^r] = [R_{ab}] = 1 \\
&[T_{mn}{}^a] = [R_{ma}] = 3/2 \\
&[R_{mn}] = 2 .
\end{aligned} \tag{52}$$

It follows that all dimension ½ tensors must vanish, since there are no covariant objects constructed from the fields in (51) with this dimension. This implies the constraints

$$T_{am}{}^n = T_{ab}{}^c = 0 . \tag{53}$$

As we shall see, the constraint $T_{am}{}^n = 0$ follows from the constraints in (47). The dimension zero tensor must be equal to a Lorentz-invariant dimensionless constant tensor, hence the only possibility is

$$T_{ab}{}^m = 2i\gamma^m{}_{ab} . \tag{54}$$

Thus the constraints on $T_{bc}{}^a$ and $T_{ab}{}^m$ are unavoidable. The dimension one tensor $T_{mn}{}^r$ could in principle be proportional to the covariant curl of the vielbein

$$T_{mn}{}^r \sim e_m{}^\mu e_n{}^\nu (D_\mu e_\nu{}^r - D_\nu e_\mu{}^r) . \tag{55}$$

Putting $T_{mn}{}^r$ equal to zero means that we go to second order formalism, in which the connection $\Omega_{mn}{}^r = \Omega_m(x,\theta)\varepsilon_n{}^r$ is no longer an independent field.

This is a choice, and not necessary: presumably there exists a first-order formalism of superspace, in which the connection $\Omega_M = E_M{}^\Lambda \Omega_\Lambda$ is an independent field, but we (and everybody else) prefer the superspace formalism with dependent Ω_M.

These were the arguments to adopt as constraints

$$\boxed{T_{mn}{}^r = 0; \quad T_{ab}{}^c = 0; \quad T_{ab}{}^m = 2i\gamma^m{}_{ab}\ .} \tag{A}$$

Let us now show, as promised, that they imply the additional constraint

$$T_{am}{}^n = 0\ . \tag{56}$$

We use the Bianchi identities for this purpose. The identity which starts with $D_a T_{bc}{}^m$ reads

$$D_a T_{bc}{}^m - T_{ab}{}^N T_{Nc}{}^m - R_{abc}{}^m + \text{supercyclic} = 0\ . \tag{57}$$

Since the tangent group is the Lorentz group, $R_{abc}{}^m = 0$ (because $(J)_c{}^m = 0$), and since $D_a \gamma_{bc}{}^m = 0$, while also $T_{ab}{}^c = 0$, we are led to

$$T_{ab}{}^n T_{nc}{}^m + T_{bc}{}^n T_{na}{}^m + T_{ca}{}^n T_{nb}{}^m = 0. \tag{58}$$

These are 8 linear equations (since (abc) = 111, 112, 122 and 222) for the 8 unknown $T_{nc}{}^m$. One may ckeck, using the explicit form of $T_{am}{}^n = 2i\gamma^m{}_{ab}$, that this system has no homogeneous solutions, hence, indeed, $T_{am}{}^n = 0$. Note that this result holds whether or not we do impose $T_{mn}{}^r = 0$. The number of constraints in $T_{ab}{}^c = 0$ and $T_{ab}{}^m = 2i\gamma^m{}_{ab}$ is 12, so that we can impose <u>at most</u> two more constraints. One possibility would be to impose no other constraint, or only <u>one</u> extra constraint. We reject this possibility since we want a superspace theory which is equivalent to x-space supergravity. If one were to impose two extra constraints, one could still choose between $T_{mn}{}^r = 0$ and $T_{mn}{}^a = 0$. The constraint $T_{mn}{}^r = 0$ is of lowest dimension and consequently more easily dealt with, but one might actually prove that the alternative, $T_{mn}{}^a = 0$, would be inconsistent.

6. The constraints as relations between covariant derivatives

Since $[D_M,D_N\}$ can be expressed in terms of $T_{MN}{}^P$ and $R_{MNP}{}^Q$, while $R_{MNP}{}^Q = R_{MN}(J)_P{}^Q$, with $(J)_P{}^Q$ diagonal in bose and fermi space, one can write the constraints also directly in terms of (anti)commutators of covariant derivatives. We claim that an equivalent set of constraints is

$$\boxed{\{D_+,D_-\} = R_{+-}J; \quad D_+D_+ = i\,D_{++}; \quad D_-D_- = i\,D_{--}} \tag{B}$$

where $D_+ = \bar{D}_a$ for a=1, $D_- = \bar{D}_a$ for a=2, and we recall that $D_{ab} \equiv \gamma^m{}_{ab}D_m$ with $\gamma^m{}_{ab}$ given in (14). Since $\gamma^m{}_{ab}$ is diagonal, only D_{++} and D_{--} are nonzero, but D_{+-} and D_{-+} are zero by definition.

To prove the equivalent of the set (B) to the set (A), we first show that (B) implies (A), and then that (A) implies (B).

If (B) holds, then from $\{D_+,D_-\} = R_{+-}J$ it follows that $T_{+-}{}^M = 0$ for any M. Hence

$$T_{+-}{}^m = T_{+-}{}^c = 0\ . \tag{59}$$

From $\{D_+,D_+\} = 2iD_{++}$ it follows that

$$T_{++}{}^a = T_{++}{}^{--} = R_{++} = 0\ ; \qquad T_{++}{}^{++} = 2i\ , \tag{60}$$

while from $D_-D_- = 2iD_{--}$ it follows that

$$T_{--}{}^a = T_{--}{}^{++} = R'_{--} = 0\ ; \qquad T_{--}{}^{--} = 2i\ . \tag{61}$$

Hence, in particular we find that

$$T_{ab}{}^c = 0, \quad T_{ab}{}^m = 2i\gamma^m{}_{ab}\ . \tag{62}$$

(because $T_{ab}{}^mD_m = 2iD_{ab}$). To show that also $T_{mn}{}^r = 0$ we consider $[D_{++},D_{--}]$. Using that $D_{++} = -iD_+D_+$ we find

$$[D_{++},D_{--}] = -[D_+D_+,D_-D_-] =$$

$$-[D_+,\{D_+,D_-\}]D_- - D_-[D_+,\{D_+,D_-\}] = -\{[D_+,\{D_+,D_-\}],D_-\}\ . \tag{63}$$

On the right-hand side one never finds a derivative D_{++} (for example, $[D_+,R_{+-}J],D_-\}$ contains $(D_+R_{+-})[J,D_-] = (D_+R_{+-})(-\frac{1}{2}D_-)$ or terms with J but never the product D_+D_+ or D_-D_-). Hence

$$T_{++,--}{}^{++} = 0 \Longleftrightarrow T_{mn}{}^{r} = 0 \ . \tag{64}$$

To prove the converse, namely that (A) implies (B), we note that from $T_{ab}{}^{c} = 0$ and $T_{ab}{}^{m} = 2i\gamma^m{}_{ab}$ it follows that

$$\{D_+,D_-\} = R_{+-}J; \quad \{D_\pm,D_\pm\} = 2i\gamma^m{}_{\pm\pm}D_m + R_{\pm\pm}J \ . \tag{65}$$

Since $\gamma^m{}_{++}D_m = D_{++}$ and $\gamma^m{}_{--}D_m = D_{--}$, all that is left to show is that $R_{++} = R_{--} = 0$. From $T_{mn}{}^{r} = 0$ it follows that

$$[D_{++},D_{--}] = -[D_+D_+ - \tfrac{1}{2}R_{++}J, \ D_-D_- - \tfrac{1}{2}R_{--}J] \ . \tag{66}$$

In this commutator one never should find terms with D_{++} or D_{--} because $T_{mn}{}^{r} = 0$. We already showed that in $[D_+D_+,D_-D_-]$ one does not find D_{++} or D_{--}. To complete the proof we note that the terms with D_{++} or D_{--} in the commutator appear as

$$-\tfrac{1}{2}R_{++}[J,D_-D_-] = \tfrac{1}{2}R_{++}D_-D_- \tag{67}$$

and idem for D_+D_+. Hence, indeed $R_{++} = R_{--} = 0$.

The set (A) is sometimes more useful than the set (B), but when we will explicitly solve the constraints, the set (B) will be more advantageous. We will first choose a convenient gauge to fix all 5×4 local x-space symmetries, the so-called superconformal gauge. Then the following result holds

(i) in the superconformal gauge, all components of the supervielbein and connection depend on only one arbitrary scalar superfield ψ .

Further, without choosing any gauge,

(ii) the covariant objects, namely supertorsions and supercurvatures, depend on only one superfield, namely R_{+-}.

(iii) the supervielbein and superconnection depend on 6 unconstrained prepotentials.

Counting shows that (i) and (iii) are consistent: making a general gauge transformation on the supervielbein in the superconformal gauge, will produce the general supervielbein which no longer is in a particular gauge (but still satisfies the constraints) and which depends on $1+5=6$ scalar superfields.

7. The superconformal gauge

We begin by evaluating the supervielbein on a basis of covariant derivatives of the coset manifold $(\mathring{D}_M)$ rather than on a basis of holonomic derivatives ∂_Λ. Hence

$$D_M = (E_M{}^\Lambda \partial_\Lambda + \Omega_M J) = \mathcal{E}_M{}^{A}\mathring{D}_A + \phi_M J. \tag{68}$$

Moreover, $\Omega_M = \phi_M$ because $\mathring{D}_A$ contains no superconnection $(\mathring{\Omega}_\Lambda = 0)$.

The geometry of the rigid coset manifold corresponds to $\mathring{\mathcal{E}}_M{}^A = \delta_M{}^A$ and $\phi_M = 0$. A priori one might expect that a conformally flat supervielbein should be of the form $e^\psi \mathring{\mathcal{E}}_M{}^A$, but since flat bose indices m scale twice as much as flat fermi indices a, one might anticipate a factor $e^{2\psi}$ in front of $\mathring{\mathcal{E}}_m{}^n$ and factor e^ψ in front of $\mathring{\mathcal{E}}_a{}^b$. Note that $\mathcal{E}_M{}^A$ connects the flat indices M of local superspace with the flat indices of rigid superspace; the latter we denote by A to avoid confusion although A = {m,a} just as M = {m,a}.

We now define the superconformal gauge by

$$\mathcal{E}_a{}^m = 0, \quad \mathcal{E}_+{}^+ = \mathcal{E}_-^- . \tag{69}$$

It is shown in the appendix that one can reach this gauge. These are indeed 5 constraints (5 superfield constraints), which fix completely all local symmetries in superspace. We shall show that in fact

$$\mathcal{E}_M{}^A = \begin{pmatrix} e^{2\psi}\delta_m{}^n & * \\ 0 & e^\psi \delta_a{}^b \end{pmatrix} \tag{70}$$

Thus, in the sectors $\mathcal{E}_a{}^b$ and $\mathcal{E}_m{}^n$ the supervielbein is the expected conformal rescaling of the coset vielbeins, but note that there are extra terms in $\mathcal{E}_m{}^b$, which depend however only on ψ.

The superconformal gauge is the superspace equivalent of the conformal gauge of the x-space vielbein, in which $e_\mu{}^m(x) = \rho(x)\delta_\mu{}^m$. For application to string theories this gauge is of particular use.

We begin by showing that as a consequence of the gauge choices in (69) we also have

$$\mathcal{E}_+{}^- = \mathcal{E}_-{}^+ = 0 . \tag{71}$$

This follows from the first constraint in set B

$$\{D_+, D_-\} = \left\{e^\psi \mathring{D}_+ + \mathcal{E}_+{}^-\mathring{D}_- + \phi_+ J,\ e^\psi \mathring{D}_- + \mathcal{E}_-{}^+\mathring{D}_+ + \phi_- J\right\} = R_{+-} J . \tag{72}$$

Since there should be no $\overset{\circ}{D}_+, \overset{\circ}{D}_-, \overset{\circ}{D}_{++}$ or $\overset{\circ}{D}_{--}$ derivatives on the right-hand side of $\{D_+, D_-\}$, in particular the coefficients of $\overset{\circ}{D}_{++}$ and $\overset{\circ}{D}_{--}$ should vanish. This immediately leads to $e^{\psi} \mathcal{E}_+{}^- = 0$ and $e^{\psi} \mathcal{E}_-{}^+ = 0$ and, hence, $\mathcal{E}_a{}^b = e^{\psi} \delta_a{}^b$.

We will now show that $\mathcal{E}_a{}^A \overset{\circ}{D}_A = e^{\psi} \overset{\circ}{D}_a$ satisfies the constraints. In the process we shall also find explicit expressions for ϕ_M and $\mathcal{E}_m{}^A$ in terms of e^{ψ}. We return to

$$\{D_+, D_-\} = \{e^{\psi} \overset{\circ}{D}_+ + \phi_+ J, \; e^{\psi} \overset{\circ}{D}_- + \phi_- J\} = R_{+-} J \, . \tag{73}$$

Evaluating the anticommutator, we find

$$e^{2\psi} \{\overset{\circ}{D}_+, \overset{\circ}{D}_-\} + e^{\psi} (\overset{\circ}{D}_+ e^{\psi}) \overset{\circ}{D}_- + e^{\psi} (\overset{\circ}{D}_- e^{\psi}) \overset{\circ}{D}_+ +$$

$$e^{\psi} (\overset{\circ}{D}_+ \phi_-) J + \phi_- e^{\psi} [J, \overset{\circ}{D}_+] +$$

$$e^{\psi} (\overset{\circ}{D}_- \phi_+) J + \phi_+ e^{\psi} [J, \overset{\circ}{D}_-] + \phi_+ [J, \phi_-] J + \phi_- [J, \phi_+] J \, . \tag{74}$$

Using that $[J, \overset{\circ}{D}_\pm] = \pm \frac{1}{2} \overset{\circ}{D}_\pm$ and $[J, \phi_\pm] = \pm \frac{1}{2} \phi_\pm$, while $\{\overset{\circ}{D}_+, \overset{\circ}{D}_-\} = 0$, we get

$$\phi_\pm = (\pm 2)(\overset{\circ}{D}_\pm e^{\psi}) \; , \quad R_{+-} = e^{\psi} (\overset{\circ}{D}_+ \phi_- + \overset{\circ}{D}_- \phi_+) - \phi_+ \phi_- \, . \tag{75}$$

This result for R_{+-} agrees with (46) since $T_{+-}{}^M = 0$.

Next we evaluate the consequences of the constraint $\{D_+, D_+\} = 2iD_{++}$. Since $D_+ = e^{\psi} \overset{\circ}{D}_+ + \phi_+ J$ we have

$$\{D_+, D_+\} = \{e^{\psi} \overset{\circ}{D}_+ + \phi_+ J, \; e^{\psi} \overset{\circ}{D}_+ + \phi_+ J\} = 2iD_{++} = 2i(\mathcal{E}_{++}{}^A \overset{\circ}{D}_A + \phi_{++} J) \, . \tag{76}$$

Evaluating the anticommutator we find

$$\overset{\circ}{D}_{++} \text{ terms: } \mathcal{E}_{++}{}^{++} = e^{2\psi}; \qquad \overset{\circ}{D}_{--} \text{ terms: } \mathcal{E}_{++}{}^{--} = 0$$

$$\overset{\circ}{D}_+ \text{ terms : } \mathcal{E}_{++}{}^+ = -2ie^{\psi} (\overset{\circ}{D}_+ e^{\psi}) \, ; \qquad \overset{\circ}{D}_- \text{ terms: } \mathcal{E}_{++}{}^- = 0$$

$$J \text{ terms : } \phi_{++} = 2e^{\psi} (\overset{\circ}{D}_{++} e^{\psi}) \, . \tag{77}$$

Similar results hold for $\{D_-, D_-\}$. Hence,

$$\mathcal{E}_M{}^A = \begin{pmatrix} e^{2\psi} \delta_m{}^n & -i\gamma_m{}^{ab} (e^{\psi} \overset{\circ}{D}_b e^{\psi}) \\ 0 & e^{\psi} \delta_a{}^b \end{pmatrix} \tag{78}$$

$$\phi_M = \{2e^{\psi} (\overset{\circ}{D}_{++} e^{\psi}); \; -2e^{\psi} (\overset{\circ}{D}_{--} e^{\psi}); \; 2(\overset{\circ}{D}_+ e^{\psi}); \; -2(\overset{\circ}{D}_- e^{\psi})\} \, . \tag{79}$$

8. Solution of the Bianchi identities

In this section we determine the most general form the supertorsions and supercurvatures can have when the geometry is constrained by the constraints in (A) or (B). Rather than inserting the constraints into the Bianchi identities, we will read off the results from set B.

(i) In the spinor-spinor sector we find the constraints

$$\{D_+, D_-\} = R_{+-}J \Rightarrow T_{+-}{}^{M} = 0 \tag{80}$$

$$\{D_\pm, D_\pm\} = 2iD_{\pm\pm} \Rightarrow R_{\pm\pm} = T_{\pm\pm}{}^{\pm} = T_{\pm\pm}{}^{\mp} = T_{\pm\pm}{}^{\mp\mp} = 0, \quad T_{\pm\pm}{}^{\pm\pm} = 2i \ . \tag{81}$$

(ii) In the spinor-vector sector we have

$$[D_\pm, D_{\pm\pm}] = [D_\pm, -iD_\pm D_\pm] = 0 \Rightarrow T_{\pm,\pm\pm}{}^{M} = R_{\pm,\pm\pm}{}^{M} = 0 \ . \tag{82}$$

Further, using $iD_{--} = D_-D_-$, we find

$$\begin{aligned}[D_+, D_{--}] &= [D_+, -iD_-D_-] = -i[\{D_+, D_-\}, D_-] \\ &= -i[R_{+-}J, D_-] = +\frac{i}{2}R_{+-}D_- + i(D_-R_{+-})J \ .\end{aligned} \tag{83}$$

It follows that

$$T_{+,--}{}^{M} = \frac{i}{2}R_{+-}\delta_-{}^{M} \ ; \quad R_{+,--} = i(D_-R_{+-}) \ . \tag{84}$$

Similarly, from $D_{++} = -iD_+D_+$ we get

$$T_{-,++}{}^{M} = -\frac{i}{2}R_{+-}\delta_+{}^{M} \ ; \quad R_{-,++} = i(D_+R_{+-}) . \tag{85}$$

(iii) Finally we consider the vector-vector sector. There is only one relation, namely $[D_{++}, D_{--}]$. One has

$$\begin{aligned}[D_{++}, D_{--}] &= -i[D_+D_+, D_{--}] = -i\{D_+, [D_+, D_{--}]\} = \\ &= -i\{D_+, \frac{i}{2}R_{+-}D_- + i(D_-R_{+-})J\} \\ &= \tfrac{1}{2}(D_+R_{+-})D_- + \tfrac{1}{2}R_{+-}R_{+-}J + (D_+D_-R_{+-})J + \tfrac{1}{2}(D_-R_{+-})D_+ \ .\end{aligned} \tag{86}$$

Hence

$$T_{++\,--}{}^{\pm} = \tfrac{1}{2}(D_{\mp}R_{+-}) ; \quad T_{++,--}{}^{\pm\pm} = 0 \tag{87}$$

$$R_{++,--} = (D_+D_-R_{+-}) + \tfrac{1}{2}(R_{+-})^2 \ . \tag{88}$$

Since we have analyzed all $[D_M, D_N\}$ we have shown that all supertorsions and supercurvatures depend only on the superfield R_{+-}, without making any gauge choice. This R_{+-} is of course a function of the 6 prepotentials. When R_{+-} vanishes, the geometry is "flat", i.e., the supertorsions take on their coset values and the supercurvatures vanish. The supervielbein is then equal to a gauge transform of the coset vielbein, because in the superconformal gauge, $R_{+-} = 0$ only if $\psi=0$ (see (93) and (79)) so that in the superconformal gauge $R_{\pm}=0$ if and only if $\&_M{}^A = \delta_M{}^A$. Hence, in a general gauge, $R_{+-} = 0$ if and only if $\&_M{}^A$ is a gauge transform of $\delta_M{}^A$.

9. General solution of the constraints

We shall now show that we can express all components of $\&_M{}^A$ and ϕ_M into the following superfields

$$\&_+{}^+, \quad \&_-{}^-, \quad \check{\&}_+{}^m, \quad \check{\&}_-{}^m, \tag{89}$$

where $\check{\&}_+{}^m$ and $\check{\&}_-{}^m$ will be defined below.

We begin by observing that the constraints $\{D_+, D_+\} = 2iD_{++}$ and $\{D_-, D_-\} = 2iD_{--}$ determine the bosonic parts of $\&_M{}^A$ and ϕ_M in terms of the fermionic parts. In detail

$$\{\&_+{}^A\overset{\circ}{D}_A + \phi_+ J, \ \&_+{}^B\overset{\circ}{D}_B + \phi_+ J\} = 2i(\&_{++}{}^A\overset{\circ}{D}_A + \phi_{++}J)$$

$$2i\&_{++}{}^A = 2\&_+{}^B(\overset{\circ}{D}_B\&_+{}^A) + 2i\&_+{}^b\&_+{}^c\gamma^m{}_{bc}\delta_m{}^A + \phi_+\&_+{}^A$$

$$2i\phi_{++} = 2\&_+{}^A\overset{\circ}{D}_A\phi_+ \ . \tag{90}$$

The remaining constraint

$$\{D_+, D_-\} = \{\&_+{}^A\overset{\circ}{D}_A + \phi_+ J, \ \&_-{}^B\overset{\circ}{D}_B + \phi_- J\} = R_{+-}J \tag{91}$$

splits into two parts

$$\{\&_+{}^A\overset{\circ}{D}_A, \ \&_-{}^B\overset{\circ}{D}_B\} - \tfrac{1}{2}\phi_+\&_-{}^A\overset{\circ}{D}_A + \tfrac{1}{2}\phi_-\&_+{}^A\overset{\circ}{D}_A = 0 \tag{92}$$

$$\&_+{}^A(\overset{\circ}{D}_A\phi_-) + \&_-{}^A(\overset{\circ}{D}_A\phi_+) - \phi_+\phi_- = R_{+-} \ . \tag{93}$$

The last equation defines R_{+-} in terms of fermionic vielbeins and connections, see (46). Hence, the only real constraint to be solved is

$$\boxed{\{\mathcal{E}_+{}^{A}\mathring{D}_A,\ \mathcal{E}_-{}^{B}\mathring{D}_B\} = \tfrac{1}{2}(\phi_+\mathcal{E}_-{}^{A} - \phi_-\mathcal{E}_+{}^{A})\mathring{D}_A\ .}$$

Let us express $\mathcal{E}_\pm{}^{A}\mathring{D}_A$ into the prepotentials and two further functions F_{++} and F_{--}, which will be fixed later, as follows

$$\begin{aligned}\mathcal{E}_+{}^{A}\mathring{D}_A &\equiv \mathcal{E}_+{}^{+}\mathring{D}_+ + \mathcal{E}_+{}^{-}\mathring{D}_- + \mathcal{E}_+{}^{m}\mathring{D}_m = \\ &= (\check{\mathcal{E}}_+{}^{+}\mathring{D}_+ + \check{\mathcal{E}}_+{}^{m}\mathring{D}_m) + F_{++}(\check{\mathcal{E}}_-{}^{-}\mathring{D}_- + \check{\mathcal{E}}_-{}^{m}\mathring{D}_m)\ .\end{aligned} \tag{94}$$

Clearly, $\mathcal{E}_+{}^{+} = \check{\mathcal{E}}_+{}^{+}$, so also $\mathcal{E}_-{}^{-} = \check{\mathcal{E}}_-{}^{-}$, while $\mathcal{E}_+{}^{-} = F_{++}\check{\mathcal{E}}_-{}^{-}$ so that $\mathcal{E}_+{}^{-}$ is determined by F_{++} (and $\mathcal{E}_-{}^{+}$ by F_{--}). Finally, $\mathcal{E}_+{}^{m} = \check{\mathcal{E}}_+{}^{m} + F_{++}\check{\mathcal{E}}_-{}^{m}$ and $\mathcal{E}_-{}^{m} = \check{\mathcal{E}}_-{}^{m} + F_{--}\check{\mathcal{E}}_+{}^{m}$ so that

$$\begin{pmatrix}\mathcal{E}_+{}^{m}\\ \mathcal{E}_-{}^{m}\end{pmatrix} = \begin{pmatrix}1 & F_{++}\\ F_{--} & 1\end{pmatrix}\begin{pmatrix}\check{\mathcal{E}}_+{}^{m}\\ \check{\mathcal{E}}_-{}^{m}\end{pmatrix} \tag{95}$$

When $F_{++}F_{--} \neq 1$ at $\theta=0$, we can express $\mathcal{E}_\pm{}^{m}$ into $\check{\mathcal{E}}_\pm{}^{m}$ and $F_{\pm\pm}$.

Let us introduce the notation

$$\check{\mathcal{E}}_+ \equiv \check{\mathcal{E}}_+{}^{+}\mathring{D}_+ + \check{\mathcal{E}}_+{}^{m}\mathring{D}_m\ ;\quad \check{\mathcal{E}}_- = \check{\mathcal{E}}_-{}^{-}\mathring{D}_- + \check{\mathcal{E}}_-{}^{m}\mathring{D}_m\ . \tag{96}$$

Then

$$\mathcal{E}_+ = \check{\mathcal{E}}_+ + F_{++}\check{\mathcal{E}}_-\ ;\quad \mathcal{E}_- = \check{\mathcal{E}}_- + F_{--}\check{\mathcal{E}}_+ \tag{97}$$

and we will now show that F_{++}, F_{--} and ϕ_+,ϕ_- can all be expressed in terms of the prepotentials

$$\mathcal{E}_+{}^{+}, \mathcal{E}_-{}^{-}, \check{\mathcal{E}}_+{}^{m}, \check{\mathcal{E}}_-{}^{m}\ . \tag{98}$$

Substituting the expansions of $\mathcal{E}_+$ and $\mathcal{E}_-$ into the constraint we obtain

$$\{\check{\mathcal{E}}_+ + F_{++}\check{\mathcal{E}}_-,\ \check{\mathcal{E}}_- + F_{--}\check{\mathcal{E}}_+\} = \tfrac{1}{2}\phi_+(\check{\mathcal{E}}_- + F_{--}\check{\mathcal{E}}_+) - \tfrac{1}{2}\phi_-(\check{\mathcal{E}}_+ + F_{++}\check{\mathcal{E}}_-)\ . \tag{99}$$

Hence

$$\begin{aligned}&(1 + F_{++}F_{--})\{\check{\mathcal{E}}_+,\check{\mathcal{E}}_-\} + (\check{\mathcal{E}}_+F_{--})\check{\mathcal{E}}_+ + (\check{\mathcal{E}}_-F_{++})\check{\mathcal{E}}_- + 2i(F_{++}\check{\mathcal{E}}_{--} + F_{--}\check{\mathcal{E}}_{++}) = \\ &= \tfrac{1}{2}(\phi_+ - \phi_-F_{++})\check{\mathcal{E}}_- - \tfrac{1}{2}(\phi_- - \phi_+F_{--})\check{\mathcal{E}}_+\ ,\end{aligned} \tag{100}$$

where $\check{\partial}_{++} = \check{\partial}_{++}{}^{A^\circ} D_A \equiv -\frac{i}{2}\{\check{\partial}_+, \check{\partial}_+\}$ and idem $\check{\partial}_{--}$.
Let us further define the "anholonomy coefficients" $\check{C}_{AB}{}^C$ by

$$\{\check{\partial}_M, \check{\partial}_N\} = \check{C}_{MN}{}^P \check{\partial}_P \ . \qquad (101)$$

Then

$$(1 + F_{++}F_{--})\check{C}_{+-}{}^{++} = -2iF_{--} \qquad (102)$$

$$(1 + F_{++}F_{--})\check{C}_{+-}{}^{--} = -2iF_{++} \qquad (103)$$

$$(1 + F_{++}F_{--})\check{C}_{+-}{}^{+} + (\check{\partial}_+ F_{--}) = -\tfrac{1}{2}(\phi_- - \phi_+ F_{--}) \qquad (104)$$

$$(1 + F_{++}F_{--})\check{C}_{+-}{}^{-} + (\check{\partial}_- F_{++}) = \tfrac{1}{2}(\phi_+ - \phi_- F_{++}) \ . \qquad (105)$$

From the first two relations we solve for F_{++} and F_{--}, while the last two yield then ϕ_+ and ϕ_-. The explicit expressions are obvious but cumbersome.

<u>Exercise</u>: Verify these results in the superconformal gauge, where $F_{++} = F_{--} = 0$ and $\phi_\pm = \pm 2\check{C}_{+-}{}^{\mp}$. Note that indeed $\check{\partial}_\pm{}^{\mp} = 0$.

10. All d=2 constraints are conventional constraints

In general there are 3 kinds of constraints: conventional constraints, representation preserving constraints and nonconformal constraints. In d=2, however, all constraints are conventional constraints. In d=4, the representation preserving constraints follow from the requirement that chiral scalar superfields can exist in local superspace. These chiral scalar superfields $\phi(x,\theta)$ are defined by $D_{\dot{A}}\phi = 0$ ($\mathring{A}$=1,2), from which as consistency conditions follows

$$\{D_{\dot{A}}, D_{\dot{B}}\}\phi = T_{\dot{A}\dot{B}}{}^M D_M \phi = T_{\dot{A}\dot{B}}{}^C D_C \phi + T_{\dot{A}\dot{B}}{}^m D_m \phi = 0 \ . \qquad (106)$$

Thus, $T_{\dot{A}\dot{B}}{}^C = T_{\dot{A}\dot{B}}{}^m = 0$, and these are the representation preserving constraints. In d=2, no chiral superfields exist for simple rigid superspace, because if $\overset{\circ}{D}_+\phi = 0$, then also $\overset{\circ}{D}_+\overset{\circ}{D}_+\phi = i\overset{\circ}{D}_{++}\phi = 0$, so that $\partial_m\phi = 0$ and ϕ would be a constant. Since in rigid superspace no chiral superfields exist, it does not make sense to attempt to restrict the local geometry (instead of ϕ) by requiring that local chiral superfields do exist. Hence in d=2 we do not find representation preserving constraints.

We shall later show that all constraints in d=2 are locally Weyl invariant, hence there are no nonconformal constraints in d=2 either. That leaves only the conventional constraints. They are conventional in the sense that one can always redefine $\mathcal{E}$ and ϕ such that after this redefinition the constraints are satisfied.

Suppose one has an arbitrary <u>unconstrained</u> supervielbein $\tilde{\mathcal{E}}_M{}^A$ and connection $\tilde\phi_M$. Then $\tilde D_M = \tilde{\mathcal{E}}_M{}^A\overset{\circ}{D}_A + \tilde\phi_M J$, and from $[\tilde D_M,\tilde D_N\}$ we find supertorsions $\tilde T_{MN}{}^P$ and curvatures $\tilde R_{MN}$ which do not satisfy the constraints, in general. Let us now show that, given $\tilde{\mathcal{E}}$ and $\tilde\phi$, one can construct from them new superfields $\mathcal{E}$ and ϕ which do satisfy the constraints. We will only show this infinitesimally: we assume that in

$$(\tilde D_\pm)^2 = i\tilde D_{\pm\pm} + \tfrac{1}{2}\delta T_{\pm\pm}{}^M\tilde D_M + \tfrac{1}{2}\delta R_{\pm\pm}J \tag{107}$$

$$\{\tilde D_+,\tilde D_-\} = (\tilde R_{+-} + \delta\tilde R_{+-})J + \delta T_{+-}{}^M\tilde D_M \tag{108}$$

all terms of order δ are small. Note that these two equations <u>define</u> the objects with δ.

Let us now discuss the construction of the new geometrical fields, $\mathcal{E}_M{}^A$ and ϕ_M, which will yield new covariant derivatives D_M which satisfy the constraints. We begin by defining $\mathcal{E}_\pm{}^M$ and $\phi_\pm$ by

$$D_\pm = \tilde D_\pm + \delta A_{\pm\pm}\tilde D_\mp + \delta C_\pm J\ . \tag{109}$$

This is not the most general parametrization (terms with $(\delta B_+{}^+)\tilde D_+$ and $(\delta B_-{}^-)\tilde D_-$ and $\tilde D_{\pm\pm}$ are lacking), but it will be sufficiently general for our purposes. Then

$$\{D_+,D_+\} = \left\{\tilde D_+ + \delta A_{++}\tilde D_- + \delta C_+J,\ \tilde D_+ + \delta A_{++}\tilde D_- + \delta C_+J\right\} =$$

$$\left\{2i\,\tilde D_{++} + \delta T_{++}{}^M\tilde D_M + \delta R_{++}J\right\} + 2(\tilde D_+\delta A_{++})\tilde D_- + 2\delta A_{++}R_{+-}J$$

$$+ 2(\tilde D_+\delta C_+)J + (\delta C_+)\tilde D_+ = 2iD_{++}\ . \tag{110}$$

Hence, this constraint provides the definition of D_{++} (of $\mathcal{E}_{++}{}^{M}$ and ϕ_{++})

$$D_{++} = \tilde{D}_{++} - \frac{i}{2}\delta T_{++}{}^{M}\tilde{D}_{M} - \frac{i}{2}\delta R_{++}J$$
$$- i(\tilde{D}_{+}\delta A_{++})\tilde{D}_{-} - \frac{i}{2}\delta C_{+}\tilde{D}_{+} - i\delta A_{++}R_{+-}J - i(\tilde{D}_{+}\delta C_{+})J \ . \tag{111}$$

A similar expression is found for D_{--}.

We must now show that $\{D_{+},D_{-}\} = R_{+-}J$. We have

$$\{\tilde{D}_{+} + \delta A_{++}\tilde{D}_{-} + \delta C_{+}J,\ \tilde{D}_{-} + \delta A_{--}\tilde{D}_{+} + \delta C_{-}J\} =$$
$$\{(\tilde{R}_{+-} + \delta\tilde{R}_{+-})J + \delta T_{+-}{}^{M}\tilde{D}_{M}\} + (\tilde{D}_{+}\delta A_{--})\tilde{D}_{+}$$
$$+ \delta A_{--}2i\tilde{D}_{++} + (\tilde{D}_{+}\delta C_{-})J + \tfrac{1}{2}\delta C_{-}\tilde{D}_{+}$$
$$+ (\tilde{D}_{-}\delta A_{++})\tilde{D}_{-} + \delta A_{++}2i\tilde{D}_{--} + (\tilde{D}_{-}\delta C_{+})J - \tfrac{1}{2}\delta C_{+}\tilde{D}_{-} \ . \tag{112}$$

This should be equal to $R_{+-}J$. Hence, we find the conditions

$$\delta T_{+-}{}^{+} + (\tilde{D}_{+}\delta A_{--}) + \tfrac{1}{2}\delta C_{-} = 0 \tag{113}$$

$$\delta T_{+-}{}^{-} + (\tilde{D}_{-}\delta A_{++}) - \tfrac{1}{2}\delta C_{+} = 0 \tag{114}$$

$$\delta T_{+-}{}^{++} + 2i\delta A_{--} = 0 \tag{115}$$

$$\delta T_{+-}{}^{--} + 2i\delta A_{++} = 0 \tag{116}$$

$$\tilde{R}_{+-} + \delta\tilde{R}_{+-} + (\tilde{D}_{+}\delta C_{-}) + (\tilde{D}_{-}\delta C_{+}) = R_{+-} \ . \tag{117}$$

From these equations one can determine

$$\delta A_{++},\ \delta A_{--},\ \delta C_{+},\ \delta C_{-} \ . \tag{118}$$

Furthermore, once (113)-(116) are satisfied, (117) becomes equal to $\{D_{+},D_{-}\} = R_{+-}J$ with R_{+-} indeed given by $\partial_{+}\phi_{-} + \dots$. Hence, the new covariant derivatives, and hence the new superfields $\mathcal{E}$ and ϕ, satisfy the constraints. One has in some sense "projected" the arbitrary $\tilde{\mathcal{E}}$ and $\tilde{\phi}$ onto $\mathcal{E}$ and ϕ which satisfy the constraints, while the remaining components can be considered to be "matter". If one drops this matter and retains only $\mathcal{E}$ and ϕ, one gets a physically different theory, but retaining everything, the new "basis" is equivalent to the old basis. In this sense conventional constraints are really not constraints but rather a basis choice for $\mathcal{E}$ and ϕ.

11. Gauge completion

One can translate the x-space theory into superspace provided the local x-space gauge algebra closes (the method of "gauge completion"). For the N=1 d=2 theory, this means that one must know the auxiliary fields. The simplest set of auxiliary fields consists of only one scalar field S(x) and the local supersymmetry transformation rules read in this case

$$\begin{aligned} e_\mu{}^m &= \tfrac{1}{2}\bar{\varepsilon}\gamma^m\psi_\mu \\ \delta\psi_\mu{}^a &= (D_\mu\varepsilon)^a + \tfrac{1}{4}\gamma_\mu\varepsilon S \\ \delta S &= \bar{\varepsilon}\tau^{\mu\nu}D_\mu\psi_\nu - \tfrac{1}{4}\bar{\varepsilon}\gamma\cdot\psi S , \end{aligned} \tag{119}$$

where

$$\begin{aligned} D_\mu\varepsilon &= \partial_\mu\varepsilon + \tfrac{1}{4}\hat{\omega}_\mu{}^{mn}\gamma_{mn}\varepsilon \quad \text{and} \\ \hat{\omega}_\mu{}^{mn} &= \omega_\mu{}^{mn}(e) + \tfrac{1}{4}(\bar{\psi}_\mu\gamma^m\psi^n - \bar{\psi}_\mu\gamma^n\psi^m + \bar{\psi}^m\gamma_\mu\psi^n) . \end{aligned} \tag{120}$$

There is no gauge action, so we cannot check the correctness of these rules by verifying the invariance of the action, but one can directly verify that the local gauge algebra closes. In particular

$$\begin{aligned} [\delta(\varepsilon_1),\delta(\varepsilon_2)] &= \delta_{gc}(\tfrac{1}{2}\bar{\varepsilon}_2\gamma^\mu\varepsilon_1) + \delta_{sup}(-\tfrac{1}{2}\bar{\varepsilon}_2\gamma^\mu\varepsilon_1\psi_\mu) \\ &+ \delta_{\ell.L.}(\tfrac{1}{2}\bar{\varepsilon}_2\gamma^\mu\varepsilon_1\hat{\omega}_\mu{}^{mn} + \tfrac{1}{4}\bar{\varepsilon}_2\gamma^{mn}\varepsilon_1 S) . \end{aligned} \tag{121}$$

The gauge completion program works as follows

(i) One chooses a gauge in superspace. The simplest gauge corresponds to

$$\begin{aligned} E_\alpha{}^a &= \delta_\alpha{}^a + \theta^\beta E_{\beta\alpha}{}^a + \mathcal{O}(\theta^2) \\ E_\alpha{}^m &= 0 + \theta^\beta E_{\beta\alpha}{}^m + \mathcal{O}(\theta^2) \\ \Omega_\alpha{}^{mn} &= 0 + \theta^\theta \Omega_{\beta\alpha}{}^{mn} + \mathcal{O}(\theta^2) , \end{aligned} \tag{122}$$

where $E_{\beta\alpha}{}^M$ and $\Omega_{\beta\alpha}{}^{mn}$ are symmetric in $\beta\alpha$. This fixes all local gauge invariances in superspace except the parameters $\Xi^\Lambda(x,\theta=0)$ and $L^{mn}(x,\theta=0)$ (the same number of local symmetries as in the x-space theory).

(ii) One substitutes these results into the contraints and solves the constraints . This is a lot of work . It turns out that the only independent x-space fields which remain in the gauge (122) are

$$E_\mu{}^m(x,\theta=0); \quad E_\mu{}^a(x,\theta=0); \quad R_{+-}(x,\theta=0), \tag{123}$$

where $R_{+-}(x,\theta=0)$ appears in $\Omega_{\beta\alpha}{}^{mn} \sim (\gamma^{mn})_{\beta\alpha} R_{+-}(x,\theta=0)$ and $R_{+-}(x,\theta)$ is the superfield into which all supertorsions and curvatures can be expressed.

(iii) One finds the most general form of the local gauge parameters $\Xi^\Lambda(x,\theta)$ and $L^{mn}(x,\theta)$ which preserve the gauge chosen. One finds that there appear only the following independent x-space parameters

$$\Xi^\mu(x,\theta=0), \quad \Xi^\alpha(x,\theta=0), \quad L^{mn}(x,\theta=0). \tag{124}$$

(iv) One deduces how $E_\mu{}^m(x,\theta=0)$, $E_\mu{}^a(x,\theta=0)$ and $R_{+-}(x,\theta=0)$ transform in superspace. By comparing with the x-space theory one then identifies

$$E_\mu{}^m(x,\theta=0) = e_\mu{}^m(x); \qquad \Xi^\mu(x,\theta=0) = \xi^\mu(x) \tag{125}$$

$$E_\mu{}^a(x,\theta=0) = \psi_\mu{}^a(x); \qquad \Xi^\alpha(x,\theta=0) = \varepsilon^\alpha(x) \tag{126}$$

$$R_{+-}(x,\theta=0) = S(x); \qquad L^{mn}(x,\theta=0) = \lambda^{mn}(x) . \tag{127}$$

A simpler procedure is to start with these identifications and to require compatibility between the local algebras in x-space and in superspace . One obtains the same results. This procedure assumes from the start that the x-space theory can be mapped into superspace, something which the first approach proves (in point iv) and which this approach proves a posteriori (because one constructs order by order in θ all superspace objects from x-space objects).

One can choose other gauge choices than in (i). For example, in the superconformal gauge (where all local symmetries are fixed) one has $E_a{}^A = \delta_a{}^A e^\psi$. Leaving the x-space symmetries unfixed one has to order θ.

$$E_a{}^\mu = 0 + h(\bar\theta\gamma^\mu)_a - he^{-\frac{1}{2}}(\bar\theta\bar\gamma^\mu)_a \tag{128}$$

$$E_a{}^\beta = e^P\delta_a{}^\beta + v(\bar\theta\gamma\cdot\psi)\delta_a{}^\beta + w\gamma^n{}_a{}^\beta(\bar\psi_m\gamma_n\gamma^m\theta) \tag{129}$$

because in the superconformal gauge in x-space ($e_\mu{}^m = e^{\frac{1}{2}}\delta_\mu{}^m$ and $\psi_\mu = \frac{1}{2}\gamma_\mu\gamma\cdot\psi$) the h and w terms vanish. In fact, from

$$E_\pm = e^\psi \mathring{D}_\pm \;; \qquad E_{++} = e^{2\psi}\partial_{\pm\pm} - 2ie^\psi(\mathring{D}_\pm\psi)\mathring{D}_\pm \tag{130}$$

we deduce that $e^{2\psi} = (\det e_\mu{}^m)^{-\frac{1}{2}} \equiv e^{-\frac{1}{2}}$, and thus $E_a{}^\beta(\theta=0) = \delta_a{}^\beta e^{-\frac{1}{4}}$ and $E_a{}^\mu(\theta=0) = 0$. Moreover,

$$E_m{}^\mu(\theta=0) = e_m{}^\mu; \quad E_m{}^\alpha(\theta=0) \sim e^{-\frac{1}{4}}\psi_m{}^\alpha \tag{131}$$

(In principle, one should allow $E_m{}^\mu(\theta=0) = e_m{}^\mu e^q$, but, as it turns out, $q=0$).

For the superparameters one can make a similar ansatz. Since the dimensions are given by

$$[\xi^\mu] = -1, \quad [\varepsilon^\alpha] = -\tfrac{1}{2}, \quad [\lambda^{mn}] = 0 \tag{132}$$

$$[e_\mu{}^m] = 0, \quad [\psi_\mu{}^a] = \tfrac{1}{2}, \quad [S] = 1 \tag{133}$$

one can make the ansatz (see appendix)

$$\mathcal{X}^\mu(\theta=0) = \xi^\mu e^q, \quad \mathcal{X}^\alpha(\theta=0) = \varepsilon^\alpha e^p + \xi\cdot\psi e^s$$

$$\Lambda^{mn}(\theta=0) = \lambda^{mn}e^r + u(\bar{\varepsilon}\gamma_5\gamma\cdot\psi)\varepsilon^{mn}. \tag{134}$$

Compatibility now requires that $q=0$ but $p=-\frac{1}{4}$. One finds in this way for example

$$(\text{s det } E)^{-1} = e^{-\frac{1}{2}} - e^{-\frac{1}{4}}\,\bar{\theta}\gamma\cdot\psi + \frac{1}{16}\,S\,\bar{\theta}\theta$$
$$+ \frac{1}{32}\,(\bar{\theta}\gamma\cdot\psi)(\bar{\theta}\gamma\cdot\psi) + \frac{1}{64}\,(\bar{\psi}_m\gamma^n\gamma^m\psi_n)\bar{\theta}\theta. \tag{135}$$

One can thus map the x-space theory in more than one way into superspace.

12. Dynamics

A gauge action of pure supergravity in x-space starts with the Einstein-Hilbert action $\kappa^{-2}(\det e)R$, where κ is the gravitational constant and has dimension $\frac{1}{2}(2-d)$ in d-dimensions (because $g_{\mu\nu} = \eta_{\mu\nu} + \kappa h_{\mu\nu}$ and $(\partial h)^2$ has dimension d). Hence, in d=2, κ is dimensionless. A superspace action must thus be of the form

$$I = \int d^2x d^2\theta\, X \tag{136}$$

where the superfield X has dimension one (x has dimension -1 and θ has dimension $-\frac{1}{2}$, and thus $d\theta$ has dimension $+\frac{1}{2}$ because $\int d\theta\,\theta = 1$. Note that the $d\theta$ in one-forms $E^M = dx^\mu E_\mu{}^M + d\theta^\alpha E_\alpha{}^M$ has dimension $-\frac{1}{2}$, and is thus a different object than the $d\theta$ of integration).

The dimensions of supertorsions and supercurvatures follow from the fact that $E_\mu{}^m(x,\theta)$ has dimension zero, and thus $E_\mu{}^a$ has dimension $\frac{1}{2}$. (Since we have nonlinear expressions in $E_\mu{}^m$ from sdet E, $E_\mu{}^m$ must be dimensionless). Recalling

$$T_{MN}{}^{P} = E_M{}^{\Lambda}(\partial_\Lambda E_N{}^{\Pi})E_\Pi{}^{P} + E_M{}^{\Lambda}\Omega_{\Lambda N}{}^{P} - (-)^{MN}(M\leftrightarrow N) \tag{137}$$

and taking Λ and Π to be bosonic for convenience one finds the following dimensions, already quoted in (52),

$$\begin{aligned} &[T_{FF}{}^{B}] = 0; \quad T_{FF}{}^{F} = T_{FB}{}^{B} = \tfrac{1}{2} \\ &[T_{BF}{}^{F}] = [T_{BB}{}^{B}] = [R_{FF}{}^{mn}] = 1 \\ &[T_{BB}{}^{F}] = [R_{FB}{}^{mn}] = 3/2; \quad [R_{BB}{}^{mn}] = 2\ . \end{aligned} \tag{138}$$

The only Lorentz-<u>scalars</u> with dimension 1 are

$$R_{\alpha\beta}{}^{mn}(\gamma_{mn})^{\alpha\beta} \simeq R_{\alpha\beta}(\tau_3)^{\alpha\beta} = R_{+-}\ , \tag{139}$$

and $T_{ma}{}^{b}(\gamma^m)^a{}_b$. However, $T_{--,+}{}^{-} = \frac{-i}{2} R_{+-}$, see (84). Moreover,

$$\int d^2x d^2\theta (\text{sdet } E) R_{+-} \tag{140}$$

is a total derivative, as we now show, and hence there is no gauge action in superspace (just as the Einstein-Hilbert action $\int eR\, d^2x$ in x-space vanishes).

To prove that $\int d^2x d^2\theta$ sdet E R_{+-} vanishes, we use the integration-by-parts theorem. Namely, as a consequence (and, to some extent, a justification) of the constraints adopted, the following theorem can be proven.

<u>Theorem</u>: $\int d^2x d^2\theta$ sdet $E(D_M v^M)(-)^M = 0$ for <u>any</u> v^M. (141)

<u>Proof</u>: We begin by noting that under a general supercoordinate transformation

$$\delta(\text{sdet } E) = \partial_\Lambda(\xi^\Lambda \text{ sdet } E)(-)^\Lambda\ . \tag{142}$$

Introducing $\xi^M \equiv \xi^\Lambda E_\Lambda{}^M$, we find that

$$\begin{aligned} \delta(\text{sdet } E) &= \partial_\Lambda(\xi^M E_M{}^\Lambda \text{ sdet } E)(-)^\Lambda = \\ &= \text{sdet } E \Big[E_M{}^\Lambda \partial_\Lambda \xi^M (-)^M + \xi^M \partial_\Lambda E_M{}^\Lambda (-)^{(M+1)\Lambda} \\ &\quad - \xi^M E_M{}^\Lambda (\partial_\Lambda E_N{}^\Pi) E_\Pi{}^N (-)^N \Big]\ . \end{aligned} \tag{143}$$

We may complete the ordinary derivatives to covariant derivatives

$$\delta(\text{sdet } E) = \text{sdet } E \Big[(D_M \xi^M)(-)^M$$

$$- (-)^N \xi^M \left(E_M{}^\Lambda D_\Lambda E_N{}^\Pi E_\Pi{}^N - E_N{}^\Lambda (D_\Lambda E_M{}^\Pi) E_\Pi{}^N (-)^{MN} \right) \Big] \qquad (144)$$

because the connection terms cancel, due to the fact that the Lorentz generators are traceless in bose and fermi space: $\Omega_{MN}{}^N(-)^N = 0$. We recognize the last terms as the trace of a supertorsion. Hence

$$\delta(\text{sdet } E) = \text{sdet } E[(D_M \xi^M)(-)^M - \xi^M T_{MN}{}^N (-)^N] \, . \qquad (145)$$

We now show that $T_{MN}{}^N$ vanishes as a consequence of the constraints we imposed. Indeed,

(i) $T_{mr}{}^r = 0$ due to $T_{mn}{}^r = 0$

(ii) $T_{ma}{}^a = 0$ since $T_{++,+}{}^+ = T_{++,-}{}^- = 0$, see the section on Bianchi identities

(iii) $T_{ar}{}^r = 0$ since $T_{ar}{}^s = 0$

(iv) $T_{ab}{}^b = 0$ since $T_{ab}{}^c = 0$.

Since $\int d^2x d^2\theta \; \partial_\Lambda v^\Lambda (-)^\Lambda = 0$ for any v^Λ, it follows that the theorem is true.

Consider now

$$R_{+-} = E_+{}^\Lambda E_-{}^\Pi (\partial_\Lambda \Omega_\Pi (-)^{\Lambda\Pi} - \partial_\Pi \Omega_\Lambda)(-)^\Lambda \; . \qquad (146)$$

We can complete the ordinary derivatives to covariant derivatives , see (46)

$$R_{+-} = D_+ \Omega_- + D_- \Omega_+ \; . \qquad (147)$$

Defining $v^M = (0, \Omega^+, -\Omega^-)$, we can apply the theorem, and see that sdet E R_{+-} is a total derivative.

Although no gauge action exists in N=1 d=2 superspace, we can still couple matter to the geometry of superspace. Consider a scalar superfield $S = A + \bar\theta\lambda + \bar\theta\theta F$. If $A(x)$ is a scalar x-space field, its x-space action reads $\int d^2x(\partial_\mu A)^2$, and hence A has dimension zero. Hence, also $[S] = 0$, and the only action which is dimensionless, and a Lorentz and general-coordinate scalar is

$$I = \int d^2xd^2\theta \quad \text{sdet}\, E\, (D_aS)(D_bS)\varepsilon^{ab} \ . \tag{148}$$

(We recall that the charge conjugation matrix is $C_{ab} = \varepsilon_{ab}$ and $\varepsilon_{ab} = \varepsilon^{ab} = +1$ for a=1,b=2). Because we may partially integrate due to the theorem derived before, the S field equation reads

$$D_aD^aS = 0 \ . \tag{149}$$

For string application one also needs the field equation of the superspace geometrical fields. However, one cannot vary $E_\Lambda{}^M$ arbitrarily to derive these field equations because the geometry is constrained. We must vary under the condition of constraints. Geometrically: one has a hypersurface in superspace, and must vary the supervielbeins such that one remains on the surface. In principle we could use our explicit solution of $\mathcal{E}_M{}^A$ (and hence $E_M{}^\Lambda$) in terms of prepotentials, since the latter are per definition the unconstrained variables. However, this procedure is tedious, (due to sdet E being nonlinear in prepotentials) and an indirect procedure yields rather easily the desired result.

For <u>any</u> variation of the vielbein $E_\Lambda{}^M$ we have

$$\delta\mathcal{L} = E_M{}^\Lambda\delta E_\Lambda{}^M(-)^M\mathcal{L} - 2E_a{}^\Lambda\delta E_\Lambda{}^M(D_MS)(D^aS) \tag{150}$$

since $\delta E_a{}^\Lambda = -E_a{}^\Pi\delta E_\Pi{}^M E_M{}^\Lambda$. Define now

$$H_M{}^N = E_M{}^\Lambda\delta E_\Lambda{}^N. \tag{151}$$

Clearly, we can extract $\delta E_\Lambda{}^N$ from $H_M{}^N$ by multiplication with $E_\Lambda{}^M$ from the left. In terms of $H_M{}^N$ one has

$$\begin{aligned}\delta\mathcal{L} &= H_M{}^M(-)^M\mathcal{L} - 2H_a{}^M(D_MS)(D^aS) \\ &= (H_m{}^m - H_a{}^a)\mathcal{L} - 2H_a{}^b(D_bS)(D^aS) - 2H_a{}^m(D_mS)(D^aS) \ .\end{aligned} \tag{152}$$

Now, $(D_bS)(D_aS) = \tfrac{1}{2}\varepsilon_{ba}\varepsilon^{cd}(D_cS)(D_dS)$. Hence

$$\delta\mathcal{L} = (H_m{}^m - 2H_a{}^a)\mathcal{L} - 2H_a{}^m(D_mS)(D^aS). \tag{153}$$

We shall now determine which variations $H_M{}^N$ respect the constraints. A priori one expects that there are six superfields which parametrize $H_M{}^N$ since there were 6 prepotentials. In other words, the hypersurface in superspace on which the constraints are satisfied, is 6-dimensional, hence there ought to be 6 directions $H_M{}^N$ in which one remains on the hypersurface. Using (137), one has in general

$$\begin{aligned}\delta T_{MN}{}^P &= \delta\Big[E_M{}^\Lambda(\partial_\Lambda E_N{}^\Pi E_\Pi{}^P + \Omega_{\Lambda N}{}^P) - (-)^{MN}(M\leftrightarrow N)\Big] \\ &= (- E_M{}^\Sigma \delta E_\Sigma{}^S E_S{}^\Lambda)(D_\Lambda E_N{}^\Pi) E_\Pi{}^P \\ &\quad - E_M{}^\Lambda \partial_\Lambda (E_N{}^\Sigma \delta E_\Sigma{}^R E_R{}^\Pi) E_\Pi{}^P + E_M{}^\Lambda \partial_\Lambda E_N{}^\Pi \delta E_\Pi{}^P + E_M{}^\Lambda \delta\Omega_{\Lambda N}{}^P - (-)^{MN}(M\leftrightarrow N) .\end{aligned} \tag{154}$$

Hence,

$$\begin{aligned}\delta T_{MN}{}^P &= - H_M{}^S T_{SN}{}^P + H_N{}^S T_{SM}{}^P (-)^{MN} - D_M H_N{}^P + D_N H_M{}^P (-)^{MN} \\ &\quad + T_{MN}{}^S H_S{}^P + E_M{}^\Lambda \delta\Omega_{\Lambda N}{}^P - E_N{}^\Lambda \delta\Omega_{\Lambda M}{}^P (-)^{MN} .\end{aligned} \tag{155}$$

Consider now $T_{ab}{}^m$. Its variation is given by

$$\delta T_{ab}{}^m = -H_a{}^S T_{Sb}{}^m - H_b{}^S T_{Sa}{}^m - D_a H_b{}^m - D_b H_a{}^m + T_{ab}{}^S H_S{}^m \tag{156}$$

because $\delta\Omega_{\Lambda b}{}^m = 0$ since $(J)_b{}^m = 0$. Inserting the constraints into the right-hand side yields the following conditions for the allowed variations

$$0 = -H_a{}^c T_{cb}{}^m - H_b{}^c T_{ca}{}^m - D_a H_b{}^m - D_b H_a{}^m + T_{ab}{}^n H_n{}^m . \tag{157}$$

We contract with $\gamma^{ab}{}_t$ to isolate $H_n{}^m$ and find

$$H_t{}^m = \frac{i}{2}\gamma^{ab}{}_t (H_a{}^c 2i\gamma^m{}_{cb} + D_a H_b{}^m) . \tag{158}$$

Tracing over t and m tells us that

$$(H_m{}^m - 2H_a{}^a) = \frac{i}{2}\gamma^{ab}{}_m (D_a H_b{}^m) . \tag{159}$$

Inserting this result into the expression for the variation of the action yields

$$\delta\mathcal{L} = \frac{i}{2}\gamma^{ab}{}_m(D_a H_b{}^m)\mathcal{L} - 2H_a{}^m(D_m S)(D^a S) \ . \tag{160}$$

After partial integration

$$\delta\mathcal{L} = H_a{}^m\left(\frac{i}{2}\gamma^{ab}{}_m D_b\mathcal{L} - 2D_m S D^a S \text{ sdet } E\right). \tag{161}$$

Since $D_a[(D_c S)(D^c S)] = 2(D_a D_c S)(D^c S)$ and $(D_a D_c S) = i\gamma^n{}_{ac}(D_n S) + \tfrac{1}{2}\varepsilon_{ac}(D_b D^b S)$, we find on-shell (where $D^2 S = 0$), that

$$\begin{aligned}\delta\mathcal{L} &= H_a{}^m(-\gamma^{ab}{}_m\gamma^n{}_{bc} - 2\delta_m{}^n\delta_c{}^a)(D_n S)(D^c S)\\ &= H_a{}^m[+(\gamma_m\gamma^n)^a{}_c - 2\delta_m{}^n\delta_c{}^a](D_n S)(D^c S)\\ &= H_a{}^m[-(\gamma^n\gamma_m)^a{}_c](D_n S)(D^c S) \ .\end{aligned} \tag{162}$$

We will shortly show that one may vary $H_a{}^m$ arbitrarily (in other words, $H_a{}^m$ are 4 of the 6 prepotentials for $H_M{}^N$). Anticipating this result, we find that the field equations of the supergravitational superfields (which are the superVirasoro constraints of string theory) are given by

$$V^{ma} \equiv (\gamma^n\gamma^m)^{ab}(D_n S)(D_b S) = 0 \ . \tag{163}$$

Let us now show that the $H_a{}^m$ are indeed unconstrained. To this purpose we analyze $\delta T_{ab}{}^c$, $\delta T_{mn}{}^r$ and the remaining parts of $\delta T_{ab}{}^m$ (the parts which are not contracted with $\gamma^{ab}{}_m$).

(i)
$$\begin{aligned}\delta T_{ab}{}^c &= -D_a H_b{}^c - D_b H_a{}^c + 2i\gamma^m{}_{ab}H_m{}^c\\ &\quad - H_a{}^m T_{mb}{}^c - H_b{}^m T_{ma}{}^c + E_a{}^{\Pi}\delta\Omega_{\Pi b}{}^c + E_b{}^{\Pi}\delta\Omega_{\Pi a}{}^c \ .\end{aligned} \tag{164}$$

Since $\Omega_{\Pi b}{}^c = \Omega_\Pi (J)_b{}^c$ and $(J)_b{}^c = (\tfrac{1}{2}\tau_3)_b{}^c$, contraction over b=c yields $E_a^{\Pi}\delta\Omega_\Pi$ in terms of $H_M{}^N$. Contraction with $\gamma^{ab}{}_n$ yields $H_n{}^c$ in terms of $\gamma^{ab}{}_n D_a H_b{}^c$.

(ii)
$$\delta T_{mn}{}^r = -D_m H_n{}^r + D_n H_m{}^r + T_{mn}{}^a H_a{}^r + E_m{}^{\Pi}\delta\Omega_{\Pi n}{}^r - E_n{}^{\Pi}\delta\Omega_{\Pi m}{}^r \ . \tag{165}$$

Contraction over n,r yields $E_n{}^{\Pi}\delta\Omega_\Pi$ in terms of $H_M{}^N$.

(iii) Contracting $\delta T_{ab}{}^{m}$ with $(\tau_3)^{ab} = \begin{pmatrix} 0 & -1 \\ -1 & 0 \end{pmatrix}$ we find

$$\tau_3{}^{ab}[- H_a{}^c\, 2i\gamma^m{}_{cb} - D_a H_b{}^m + i\gamma^n{}_{ab} H_n{}^m] = 0 \ . \tag{166}$$

With $(\tau_3\gamma^n)^a{}_a = 0$ it follows that

$(\tau_3\gamma^m)^a{}_c H_a{}^c$ can be expressed in terms of $H_b{}^m$.

Since $\tau_3\gamma^m = \gamma^{1-m}$ for m=0,1, it follows that $H_a{}^a$ and $(\tau_3 H)_a{}^a$ are unconstrained.

The conclusion is therefore that all variations $H_M{}^N$ can be expressed in terms of $H_a{}^m$, $H_a{}^a$ and $(\tau_3 H)_a{}^a$, or, equivalently, in terms of

$$H_a{}^m,\ H_+{}^+ \text{ and } H_-{}^- \ . \tag{167}$$

13. Local Weyl invariance

We define local Weyl transformations as those transformations which leave constraints invariant and under which the fermionic inverse vielbein rescales as

$$\mathcal{E}_+{}^A \rightarrow \Lambda^{\frac{1}{2}}(x,\theta)\mathcal{E}_+{}^A \tag{168}$$

or, equivalently,

$$E_+{}^\Lambda \rightarrow \Lambda^{\frac{1}{2}}(x,\theta)E_+{}^\Lambda \ . \tag{169}$$

A priori it is not obvious that such transformations exist. In d=4, there are some constraints which are not Weyl invariant, but requiring that in d=2 all constraints are Weyl invariant means that we select a geometry suitable for a locally superconformally invariant theory. It is also not obvious that all components of $\mathcal{E}_+{}^A$ should scale in the same way, although that is a natural requirement because Weyl transformations act on the index M of $\mathcal{E}_M{}^A$. One might expect that $\mathcal{E}_{++}{}^A \rightarrow \Lambda(x,\theta)\mathcal{E}_{++}{}^A$ under local Weyl rescalings, but that does not leave the constraints invariant as we shall see.

The above transformation rules contain the information that the pre-potentials $\mathcal{E}_+{}^+$, $\mathcal{E}_+{}^m$, $\mathcal{E}_-{}^-$ and $\mathcal{E}_-{}^m$ all scale with $\Lambda^{\frac{1}{2}}$. Indeed, recalling that $\mathcal{E}_+{}^- = F_{++}\check{\mathcal{E}}_-{}^-$, it follows that F_{++} (and F_{--}) are Weyl inert, and hence $\mathcal{E}_\pm{}^m$ and $\check{\mathcal{E}}_\pm{}^m$ scale in the same way.

Recalling that from $\{D_\pm, D_\pm\} = 2iD_{\pm\pm}$ we obtained

$$2i\mathcal{E}_{\pm\pm}{}^{A} = 2\mathcal{E}_{\pm}{}^{B}\mathring{D}_{B}\mathcal{E}_{\pm}{}^{A} + 2i\mathcal{E}_{\pm}{}^{b}\mathcal{E}_{\pm}{}^{c}\gamma^{m}{}_{bc}\delta^{A}{}_{m} \pm \phi_{\pm}\mathcal{E}_{\pm}{}^{A} \tag{170}$$

$$2i\,\phi_{\pm\pm} = 2\mathcal{E}_{\pm}{}^{A}\mathring{D}_{A}\phi_{\pm} \tag{171}$$

we can determine how $\mathcal{E}_{\pm\pm}{}^{A}$ and $\phi_{\pm\pm}$ transform, provided we know how $\phi_{\pm}$ transforms.

One might decide to directly solve for $\phi_{\pm} = \Omega_{\pm}$ from $T_{ab}{}^{c} = 0$. By defining

$$T_{ab;c} = C_{ab;c} + \Omega_{abc} + \Omega_{bac} , \tag{172}$$

where $C_{ab;c}$ are the Ω independent terms, one finds

$$2\Omega_{abc} = 2\Omega_{a}\begin{pmatrix} 0 & \tfrac{1}{2} \\ \tfrac{1}{2} & 0 \end{pmatrix}_{bc} = - C_{ab;c} - C_{ac;b} + C_{bc;a} . \tag{173}$$

Hence

$$\Omega_{+} = -C_{++}{}^{+} , \qquad \Omega_{-} = +C_{--}{}^{-} . \tag{174}$$

However, in

$$C_{++}{}^{+} = 2E_{+}{}^{\Lambda}(\partial_{\Lambda}E_{+}{}^{\Pi})E_{\Pi}{}^{+} \tag{175}$$

one encounters $E_{\Pi}{}^{+}$ whose transformation rule one does not know at this point. Thus, solving ϕ_{+} from $T_{ab}{}^{c} = 0$ is not a good strategy.

Rather, we go back to the constraint $T_{+-}{}^{A} = 0$ in (92)

$$\{\mathcal{E}_{+}{}^{A}\mathring{D}_{A}, \mathcal{E}_{-}{}^{B}\mathring{D}_{B}\} = \tfrac{1}{2}(\phi_{+}\mathcal{E}_{-}{}^{A} - \phi_{-}\mathcal{E}_{+}{}^{A})\mathring{D}_{A} \tag{176}$$

and see immediately how $\phi_{\pm}$ transforms

$$\phi_{\pm} \to \Lambda^{\frac{1}{2}}\phi_{\pm} \pm 2\mathcal{E}_{\pm}{}^{A}(\mathring{D}_{A}\Lambda^{\frac{1}{2}}) . \tag{177}$$

Hence

$$\mathcal{E}_{\pm\pm}{}^{A} \to \Lambda\mathcal{E}_{\pm\pm}{}^{A} - i\mathcal{E}_{\pm}{}^{B}(\mathring{D}_{B}\Lambda)\mathcal{E}_{\pm}{}^{A} . \tag{178}$$

For the bosonic connection one finds the following behaviour under Weyl transformations

$$\phi_{\pm\pm} \to \Lambda\phi_{\pm\pm} - \frac{i}{2}\mathcal{E}_{\pm}{}^{A}(\mathring{D}_{A}\Lambda)\phi_{\pm} \mp 2i\Lambda^{\frac{1}{2}}\mathcal{E}_{\pm}{}^{A}\mathring{D}_{A}(\mathcal{E}_{\pm}{}^{B}\mathring{D}_{B}\Lambda^{\frac{1}{2}}) . \tag{179}$$

As a check we note that in the superconformal gauge, the replacement $e^{\psi} \to e^{\psi}\Lambda^{\frac{1}{2}}$ preserves the superconformal gauge and corresponds to a Weyl rescaling.

For example,

$$2i\phi_{++} = 2\mathcal{E}_{+}{}^{A}\overset{\circ}{D}_{A}\phi_{+} = 2e^{\psi}\overset{\circ}{D}_{+}(2\overset{\circ}{D}_{+}e^{\psi}) = 4ie^{\psi}D_{++}e^{\psi} \tag{180}$$

and the replacement $e^{\psi} \to e^{\psi}\Lambda^{1/2}$ agrees with

$$\phi_{++} \to \Lambda\phi_{++} - \frac{i}{2}e^{\psi}(\overset{\circ}{D}_{+}\Lambda)(2\overset{\circ}{D}_{+}e^{\psi}) - 2i\Lambda^{1/2}e^{\psi}\overset{\circ}{D}_{+}(e^{\psi}\overset{\circ}{D}_{+}\Lambda^{1/2}) \tag{181}$$

as one may verify.

Since we deduced the Weyl transformation rules of the various superfields from the constraints, the constraints are automatically Weyl invariant. Hence, there are no "nonconformal constraints" in $d=2$, unlike the case of $d=4$.

It is now clear that simple $d=2$ superspace is superconformally flat. Namely, we can always transform a given supervielbein to the superconformal gauge, in which the supervielbein is just a Weyl rescaling of the rigid vielbein of the coset manifold.

14. Field equations and Virasoro constraints in x-space

Consider again the coupling of a scalar superfield to the superspace geometry

$$I = \int d^2x d^2\theta (\text{s}\det E_{\Lambda}{}^{M})\, \varepsilon^{ab}(D_a S)(D_b S)\ . \tag{182}$$

Note that $\text{sdet}\, E_{\Lambda}{}^{M} = (\text{sdet}\, E_{M}{}^{\Lambda})^{-1}$ and $\text{sdet}\, E_{M}{}^{\Lambda} = (\text{sdet}\, \mathcal{E}_{M}{}^{A})(\text{sdet}\, \overset{\circ}{E}_{A}{}^{\Lambda}) = \text{sdet}\, \mathcal{E}_{M}{}^{A}$. Since $\mathcal{E}_{\pm\pm}{}^{A}$ transforms under Weyl scaling into Λ times itself plus a linear combination of $\mathcal{E}_{\pm}{}^{A}$, we have

$$\text{sdet}\ \mathcal{E}_{M}{}^{A} \to \Lambda(x,\theta)\ \text{sdet}\ \mathcal{E}_{M}{}^{A}\ . \tag{183}$$

Recalling that $D_a S = \mathcal{E}_a{}^{A}(\overset{\circ}{D}_A S)$ transforms into $\Lambda^{1/2}$ times itself, it follows that the action is locally Weyl invariant.
Since the action is general supercoordinate and locally Lorentz invariant, we can choose a convenient gauge, the superconformal gauge. Due to the Weyl invariance, all ψ dependence cancels, and the action becomes simply

$$I = \int d^2x d^2\theta\ \varepsilon^{ab}(\mathring{D}_a S)(\mathring{D}_b S)\ . \tag{184}$$

The field equations reduce to

$$\mathring{D}_a \mathring{D}^a S = 0 \tag{185}$$

and the Virasoro constraints become

$$V^{ma} = (\gamma^n \gamma^m)^{ab}(\partial_n S)(\mathring{D}_b S) = 0\ . \tag{186}$$

Let us expand these two equations into θ's. For the field equation one finds

$$\varepsilon^{ab}(\partial_a + i\theta^c \not\partial_{ca})(\partial_b + i\theta^d \not\partial_{db})S = 0 \tag{187}$$

$$S = X + \theta^c \lambda_c + \theta^a \theta_a F \tag{188}$$

$$\theta^0:\ F = 0.\quad \text{Hence}\quad S = X + \theta^c \lambda_c \tag{189}$$

$$\theta^1:\ 2\varepsilon^{ab}(i\theta^c \not\partial_{ca}\lambda_b) = 0 \Rightarrow \not\partial\lambda = 0 \tag{190}$$

$$\theta^2:\ \varepsilon^{ab}\theta^c\theta^d \not\partial_{ca}\not\partial_{db}X = 0 \Rightarrow \square X = 0\ . \tag{191}$$

Hence, we find as field equations of the spinning string

$$\square X = \not\partial\lambda = F = 0\ . \tag{192}$$

The Virasoro constraints can similarly be expanded into θ's.

$$(\gamma^n\gamma^m)^{ab}(\partial_n S)(\partial_b S + i\theta^c \not\partial_{cb} S) = 0 \tag{193}$$

θ^0: $(\not\partial X)(\gamma^m\lambda) = 0$. These are the fermionic parts of the super-Virasoro constraints. (194)

$$\theta^1:\ (\gamma^n\gamma^m)^{ab}(\partial_n\theta^d\lambda_d)\lambda_b + (\gamma^n\gamma^m)^{ab}(\partial_n X) i\theta^d \not\partial_{db} X = 0$$

$$(\gamma^n\gamma^m\lambda)^a(\partial_n\lambda_d) - i(\gamma^n\gamma^m\gamma^t)^a{}_d(\partial_n X\partial_t X) = 0 \tag{195}$$

Tracing with $(I)^d{}_a$ or $(\tau_3)^d{}_a$ yields zero, while tracing with $(\gamma^s)^d{}_a$ yields

$$+\ (\partial_n\bar{\lambda})(\gamma^s\gamma^n\gamma^m\lambda) - i\ \mathrm{tr}(\gamma^n\gamma^m\gamma^t\gamma^s)(\partial_n X)(\partial_t X) = 0 \tag{196}$$

and since $\gamma^n\gamma^m\gamma^t\partial_n X\partial_t X = 2\partial_m X\not\partial X - \gamma^m(\partial X)^2$ we find the bosonic part of the superVirasoro constraints.

$$i(\partial_n \bar{\lambda})\gamma_m \lambda + \left(2\partial_m X \partial_n X - \eta_{mn}(\partial X)^2\right) = 0 \qquad (197)$$

(If one defines the Dirac bar by means of $\lambda^\dagger \tau_2$ instead of $\lambda^\dagger \gamma_0$, the factor i disappears).

$$\theta^2\colon\ (\gamma^n\gamma^m)^{ab}(\partial_n \theta^d \lambda_d)(i\theta^c \not{\partial}_{cb} X) + (\gamma^n\gamma^m)^{ab}\,\partial_n X\ i\theta^c \not{\partial}_{cb} \theta^d \lambda_d = 0\ . \qquad (198)$$

Using $\theta^c \theta^d = \varepsilon^{cd}\theta^2$ we find a term

$$\gamma^n\gamma^m \not{\partial} X \partial_n \lambda = 0 \qquad (199)$$

since the last term vanishes on-shell. With $\gamma^n\gamma^m\gamma^t = \gamma^t\gamma^m\gamma^n$ also the first term vanishes.

Hence, the θ-expansion reproduces the super-Virasoro constraints.

REFERENCES

Most of the work on two-dimensional superspace was done by Howe [1,2]. Similar results were obtained in [3]. A solution of the constraints in a general gauge was obtained in [4]. For an introductory set of lectures see also [5]. Superspaces results have been used for heat kernel computations in ref. [6]. Recent articles on chiral d=2 superspaces can be found in [7-9]. An alternative multiplet for scalars in rigid supersymmetry can be found in ref. [10].

[1] P.S. Howe, J. Phys. A12, 393 (1979).

[2] P.S. Howe, Phys. Lett. 70B, 453 (1977).

[3] M. Brown and S.J. Gates, Ann. Phys. 122, 443 (1979).

[4] M. Roček, P. van Nieuwenhuizen and S.C. Zang, Ann. Phys., 172, 348 (1986).

[5] M.T. Grisaru, lectures at the 1968 Trieste school on supersymmetry, supergravity and superstrings.

[6] E. Martinec, Phys. Rev. D28, 2604 (1983).

[7] S.J. Gates and H. Nishino, to appear in Class. and Quant. Gravity. S.J. Gates, M.T. Grisaru, L. Mezincescu and P.K. Townsend, UTTG-20-86 and BRX-TH-205.

[8] M. Evans and B. Ovrut, Phys. Lett. 171B, 177 (1986), 175B, 145 (1986).

[9] G. Moore and P. Nelson, Harvard preprint HUTP/86/A014.

[10] S. Ferrara, Lett. Nuov. Cim. 13, 629 (1973).

APPENDIX

In this appendix we show that one can indeed reach the superconformal gauge in (69).

The transformation properties of $E_M{}^\Lambda$ and $\Omega_M{}^i$ can be succinctly defined by

$$\delta D_M = [K, D_M], \qquad K = K^\Lambda \partial_\Lambda + L^i H_i$$

(In our case, $L^i H_i = LJ$). Indeed, upon expansion of

$$(\delta E_M{}^\Lambda)\partial_\Lambda + \delta(E_M{}^\Lambda \Omega_\Lambda{}^i)H_i = [K^\Sigma \partial_\Sigma + L^j H_j, E_M{}^\Lambda \partial_\Lambda + \Omega_M{}^i H_i]$$

one obtains

$$\delta E_M{}^\Lambda = -E_M{}^\Sigma \partial_\Sigma K^\Lambda + K^\Sigma \partial_\Sigma E_M{}^\Lambda + L_M{}^N E_N{}^\Lambda$$

$$\delta \Omega_M{}^i = K^\Sigma \partial_\Sigma \Omega_M{}^i + L_M{}^N \Omega_N{}^i - E_M{}^\Lambda \partial_\Lambda L^i - \Omega_M{}^j L^k c_{jk}{}^i$$

$$\delta \Omega_{MN}{}^P = K^\Sigma \partial_\Sigma \Omega_{MN}{}^P - D_M L_N{}^P + L_M{}^{M'} \Omega_{M'N}{}^P$$

with

$$D_M L_N{}^P = \partial_M L_N{}^P + \Omega_{MN}{}^{N'} L_{N'}{}^P - L_N{}^{P'} \Omega_{MP'}(-)^{M(N+P')} .$$

For the Lorentz group this reduces to

$$D_M L_N{}^P = (\partial_M L)(J)_N{}^P .$$

Let us now expand the supervielbein and superconnection about the geometry of the coset manifold. Then we define

$$D_M = \mathcal{E}_M{}^{A}\mathring{D}_A + \phi_M{}^i H_i \quad \text{and}$$

$$\delta D_M = \delta(\mathcal{E}_M{}^{A}\mathring{D}_A + \phi_M{}^i H_i) = [K, D_M]$$

$$K = \kappa^{A}\mathring{D}_A + \ell^i H_i .$$

It follows that then

$$\delta \mathcal{E}_M{}^A = \kappa^{B}\mathring{D}_B \mathcal{E}_M{}^A - \mathcal{E}_M{}^{B}\mathring{D}_B \kappa^A - \mathcal{E}_M{}^b \kappa^c 2i\gamma^m{}_{bc}\delta^A{}_m + \ell_M{}^N \mathcal{E}_N{}^A$$

$$\delta \phi_M{}^i = \kappa^{A}\mathring{D}_A \phi_M{}^i - \mathcal{E}_M{}^{A}\mathring{D}_A \ell^i + \ell_M{}^N \phi_N{}^i + \phi_M{}^j \ell^k c_{kj}{}^i .$$

In particular

$$\delta \mathcal{E}_\pm{}^\pm = \kappa^{B}\mathring{D}_B \mathcal{E}_\pm{}^\pm - \mathcal{E}_\pm{}^{B}\mathring{D}_B \kappa^\pm \pm \tfrac{1}{2}\ell \mathcal{E}_\pm{}^\pm$$

$$\delta \mathcal{E}_\pm{}^{\pm\pm} = \kappa^{B}\mathring{D}_B \mathcal{E}_\pm{}^{\pm\pm} - \mathcal{E}_\pm{}^{B}\mathring{D}_B \kappa^{\pm\pm} - \mathcal{E}_\pm{}^\pm \kappa^\pm (2i) \pm \tfrac{1}{2}\ell \mathcal{E}_\pm{}^{\pm\pm}$$

$$\delta \mathcal{E}_\pm{}^{\mp\mp} = \kappa^{B}\mathring{D}_B \mathcal{E}_\pm{}^{\mp\mp} - \mathcal{E}_\pm{}^{A}\mathring{D}_A \kappa^{\mp\mp} - \mathcal{E}_\pm{}^\mp \kappa^\mp (2i) \pm \tfrac{1}{2}\ell \mathcal{E}_\pm{}^{\mp\mp} .$$

Since the coset values are $\overset{\circ}{\mathcal{E}}_M{}^A = \delta_M{}^A$ and $\overset{\circ}{\phi}_M{}^i = 0$ it follows that we can reach the following five gauges:

(i) : the algebraic gauge $\mathcal{E}_+{}^+ - \mathcal{E}_-{}^- = 0$. This fixes $\mathcal{L}$.

(ii) and (iii):

the algebraic gauges $\mathcal{E}_+{}^{++} = \mathcal{E}_-{}^{--} = 0$. These fix $\mathcal{K}^+$ and $\mathcal{K}^-$.

(iv) and (v):

the differential gauge $\mathcal{E}_+{}^{--} = \mathcal{E}_-{}^{++} = 0$.
These fix $\mathcal{K}^{++}$ and $\mathcal{K}^{--}$. (Because $\mathcal{K}^{++}$ has as many degrees of freedom as $\mathcal{E}_+{}^{--}$, this is possible).

DISCUSSION I

CHAIRMAN: P. van Nieuwenhuizen

Scientific Secretaries: M. Carter, U. Kraemmer, M. McGuigan and P. Neuman

– *S. Giddings:*

In D=2 supergravity, even though you have given an argument about the necessity of imposing your torsion constraints, it still appears that they are put in by hand. Is there a possibility that they are determined by some more fundamental structure?

– *P. van Nieuwenhuizen:*

There is no way to imagine relaxing the constraints on T^m_{ab} or T^a_{bc}, there is nothing else they can be. T^m_{ab} must be a constant tensor with this index structure, and the requirement that rigid supersymmetry is contained in local supersymmetry fixes it exactly. One could imagine relaxing the constraint $T^p_{ab} = 0$. This would lead to a different theory. These constraints are indeed still imposed by hand; one may hope to obtain them in a more natural way. Perhaps from field equations of some other formulations. There have been attempts to do this from the group manifold approach but this has not worked so far. It appears that there are no better ways to get the torsion constraints.

– *M. Carter:*

A few years ago it was popular to consider theories with conformal symmetry at very high energy which was broken at low energy giving Einstein gravity with Newton's constant as the expectation value of the dilaton field. It had problems with poles in the graviton propagator which violated unitarity. How do you construct conformal supergravity and does it have the same disease?

– *P. van Nieuwenhuizen:*

There are two types of supergravities: Poincaré and Conformal. The problem of the pole being on the wrong sheet of the Riemann surface still exists in the superconformal theories. However several people believe that this problem will be solved and conformal gravity will still be the correct theory of gravity. Antoniadis and Tomboulis have put out a long paper on the unitarity problem in the past year.

In any case, you can also view string theories as conformal supergravities. In the usual approach, one writes down a Lagrangian and finds out afterwords that it has local Weyl invariance and local conformal symmetry. My view is that, in the N=1 model, this is not an accident and should be put in from the start. If you start with a superconformal algebra, this all works out simply. It gives the same theory as the Poincare approach, except that the auxillary field S, which plays no role anyhow, disappears [(1)]. For N=2, it solves a problem first observed by Fradkin and Tsetylin. They considered the N=2 string theory with a vector field, which has of course two components in two dimensions, and wanted to gauge the vector field away completely and for that they needed two local symmetries. I never understood this. In conformal supergravity, there are indeed two local symmetries and <u>two</u> gauge fields but in the action, you only get a linear combination of them. Thus one can indeed gauge away this one gauge field by the two local symmetries. However, all this is classical, and the quantum level one of the two local symmetries has an anomaly. One should then apply QCD(2) path–integral techniques, see A. Ceresole, A. Lerda, P. Pizzochero and myself, Phys. Lett. B to be published.

– *M. McGuigan:*

Might one go beyond the minimal covariant action via extended supersymmetries or some sort of non–covariant action on the world sheet?

– *P. van Nieuwenhuizen:*

Years ago there was a meeting in which Feynman asked the question can one not consider non–covariant actions whose field equations are covariant. We always write down a covariant action when we want covariant field equations. I saw recently some papers by some people from Götherberg in Sweden in which they constructed non–covariant actions for gravity whose field equations were covariant. So we could reconsider the string theory using non–covariant actions.

As for extended supersymmetries there are variety of models which have been found in two dimensions. I shall give a partial list here.

1. "N=1 Coupled to Supergravity in x–Space" due to Brink, Howe, Di Vecchia, Deser, Zumino, P.L. 1976.
2. "N=1 Coupled to Supergravity in Superspace" Howe, J. Phys. A, 1976.
3. "Rigid N=1 with Wess–Zumino Term, (Parallizing Torsion)", Curtright, Zachos, P.R.L. 1984.
4. "N=1 with Wess–Zumino Term Coupled to Supergravity", Bergshoff, Randjbar–Daemi, Salam, Sergin, Sarmedi, 1985.

---

(1) P. van Nieuwenhuizen, Journal of Modern Physics A, World Scientific Publishing Co.

5. It was shown that Rigid N=2 and N=4 are Kahler and Hyper–Kehler Respectively by: Alvarez Gaumè and Freedman, C.M.P. 1981.
6. "Rigid N=2 and N=4 Models with a Wess–Zumino Term" by Gates, Hull, Roček 1984.
7. "The N=2 Coupling to Supergravity" was done by Brink and Schwartz, N.P. 1977.
8. "The N=4 Coupling to Supergravity" was shown to be "Hyper–Kahler or Quaternionic", Pernici and myself, P.L. 1986.
9. "N=4 with Wess–Zumino Term Coupled to Supergravity" (the most general case today) was developed by de Wit and myself (Utrecht preprint).
10. "Heterotic Models which Treat Left and Right Movers Separately" have been discussed by Bergshoeff, Sezgin, Kallosh, 1986. Also Evans and Ovrut (Rockefeller preprint), Moore and Nelson (Harvard preprint).

– *F. Quevedo:*

1) Torsion constants, is that related to what Witten did in a recent paper on twistors in which he argued you could get the torsion constraints out of that?
2) In the case of pure gravity, using the Palatini formalism you can get the torsion constraint from the field equations. Cannot you do something like that in superspace?

– *P. van Nieuwenhuizen:*

1) He gets a derivation of some of the constraints, but not all of them, by means of light–cone surfaces. There is also a paper by Freund and Mezincescu on that subject.
2) No it turns out that these constraints are not field equations. Adopting these constraints or not, leads to different physical theories. In gravity the situation is different, there you can either treat the spin connection as an independent field or solve its field equation because it is algebraic and you get the same theory. It does not matter whether you use the first or second order formalism to gravity. For superspace it would make a difference and in that sense that alternative superspace theory I was mentioning might be interesting to work out.

– *A. Shapere:*

Is it conceivable that all consistent supergravity theories come from superstring theories as field theory limits of compactified versions?

– *P. van Nieuwenhuizen:*

I do not think so. For example, in 4 dimensions if I take N=4 supergravity coupled to 1867 Yang–Mills multiplets that would not come from strings. Strings have a very particular composition. In supergravity you have the freedom of adding more and more matter multiplets and that, of course, on the one hand, is the weakness of supergravity in that it has more freedom than strings. On the other hand, because you have more freedom, supergravity may still be able to agree with experiment.

– *A. Shapere:*

What is the status of the finiteness of supergravity theories?

– *P. van Nieuwenhuizen:*

There has not been much progress. It is still an open point. Most people believe that supergravity is not finite but nobody has proven that. The status is that people have lost interest because the problem is too difficult. It was a very hot problem during the first couple of years, especially when the one loop and two loop corrections turned out to be finite. The status of the three loop corrections is that you can write a possible dangerous counterterm but you would have to compute its coefficient to find out if it is there or not. If you find that it is non–zero then you would have proven that supergravity is not a good theory. I know that some people are contemplating that computation. You may know that last year some people proved that two loop gravity is not finite by an explicit computation. The three loop supergravity computation is much, much more complicated. But some people want to do that.

– *A. Shapere:*

Please, explain why the cosmological constant is zero in eleven dimensions.

– *P. van Nieuwenhuizen:*

Yes, that is an extremely simple computation which at the time we did not think it was very important. So we published it in Nuovo Cimento Letters *. I hope we are not insulting anyone. It turned out to be the end of 11 dimensional supergravity. In other dimensions (up to and including 7) you can add to the action of supergravity a cosmological constant and still maintain local supersymmetry. In eleven dimensions (and also in ten) it turns out you cannot add a cosmological constant to the action maintaining local supersymmetry. It is a very simple and straightforward computation. At the time we did not think of Kaluza–Klein supergravity. What happens is 11 dimensions splits into four and seven. Seven to be curved into a compact space and the other four open. However in four dimensional spacetime one gets an enormous cosmological constant of the same size (but with opposite sign) as in 7 dimensional compact space. They have to cancel each other

---

* M. Nicolai, P.K. Towsend and P. van Nieuwenhuizen, Nuovo Cimento Letters.

because you did not have one in eleven dimensions. I do not think that there is any way around that problem. Unless somebody finds an extra term you can add to 11 dimensions supergravity that is a kind of Chern–Simon term. Such a Chern–Simon term looks like $tr\ R\wedge R\wedge R\wedge R\wedge R\wedge R$ or $(trR\wedge R)tr(R\wedge R\wedge R\wedge R)$ etc. So far no one has found a supersymmetric extension for eleven dimensions. If one could find it in my opinion all of the $K-K$ theory for 11 dimensions supergravity would have to begin again and all of the nice features that we have learned from 10 dimensions will carry over.

– *D. Rohrlich:*

How does supergravity increase our understanding of general relativity as a gauge theory?

– *P. van Nieuwenhuizen:*

There are 2 different approaches to this question: the coset approach and the group manifold approach. These are 2 different things; in the case of coset–manifolds G/H one usually takes for H the the Lorentz group and then G would be the super Poincaré. In the group–manifold approach you just take the whole super Poincaré group.

In the case of coset–manifolds you always have 2 geometrical fields, the super–vielbein and the spin connection and they play a completely different role. The spin connection Ω is a Yang–Mills field, whereas the vielbein in this approach is not the gauge–field of the translation generator. In the case of the group manifold you have for every generator a gauge field, so you would have E and Ω, but now they are on equal footing, E corresponding to the translation generator and Ω to the Lorentz–generators. You would like to have a connection between these 2 approaches. This comes about as follows: in the group manifold approach you require that the theory is not invariant under the P–gauge transformation (Yang–Mills transformation) but it should be invariant under general coordinate transformation. This is a kind of physical input. The way to do is to put the curvature of this P–generator equal to zero.

– *D. Rohrlich:*

Does it turn out that the gauge group is the Lorentz group?

– *P. van Nieuwenhuizen:*

In the coset–approach you can take any subgroup H. The reason why we take the Lorentz group is not any theoretical reason. It is a simple experimental evidence that, if you take another subgroup than the Lorentz group you will get theories with non positive energies.

– *K. Meissner:*

You mentioned that for spins greater than two we do not have a consistent field theory. But the string theory has an infinite ladder of spins and seems to be a consistent theory. Can the same thing happen in the supergravity theory?

– *P. van Nieuwenhuizen:*

Supergravity was (up to now) constructed only for a finite number of fields and in that case no example of a consistent theory for spins greater than two was found. But no one has proved that such a theory cannot exist. I looked for a consistent theory for spins 5/2 and 2 and it seems that I exhausted all possibilities with no result(*). The string theory indeed seems to be consistent but it contains an infinite number of fields and spins. It might be possible to construct supergravity theories with an infinite number of fields and arbitrarily large spins which are consistent but not equivalent to string theories, but that is an open problem.

– *F. Quevedo:*

I want to comment on Shapere's question about the cosmological constant in D=11 supergravity. There is a paper from Duff and Orzalesi where they obtain a vanishing cosmological constant after compactification assuming fermions play a role.

– *P. van Nieuwenhuizen:*

Yes, you can add extra features in a field theory. They assumed that a fermion bilinear obtains a vacuum expectation value. However you will need to prove that this happens at the quantum level and at present not much can be said about it.

DISCUSSION II

– *M. Carter:*

In d=4 N=1 supergravity one can add a super cosmological constant of the form $\Lambda + b\overline{\Psi}_\mu \gamma^{\mu\nu} \Psi_\nu$ where b is an appropriate constant which preserves supersymmetry and leaves the gravitino massless. Why cannot I do something similar into the four dimensional reduction of a superstring theory, which includes Yang–Mills fields?

– *P. van Nieuwenhuizen:*

Nothing seems known about the possibility of obtaining gauged four–dimensional supergravity theories directly from four–dimensional string theories. The only thing that is known is that one can first obtain ten–dimensional supergravities (which never have a super–cosmological constant) from ten–dimensional string theories, and then obtain after a Kaluza–Klein compactification on a suitable compact six–dimensional surface a four–dimensional supergravity with super cosmological constant. In string theories the "easiest" problem is to obtain the spectrum of particles, but this spectrum, but this spectrum is the same for theories with or without super cosmological constant.

– *Y. Shamir:*

Why is it that in supergravity theories you do not have the same trivial relation between supersymmetry generators and the Hamitonian that implies that the vacuum energy is zero?

– *P. van Nieuwenhuizen:*

In supergravity, as in any gauge theory, there is no local charge. In fact, the energy E and the supersymmetry charge Q are in supergravity surface integrals, and are so–called first–class constraints, whose bracket relations are more complicated than $\{Q, Q\} = P$. Witten has given a proof that $E \geq 0$ for anti–de Sitter spacetimes as well as flat space–times by replacing Q by some differential operator D_μ, and using that the energy can be expressed in terms of the commutator $[D_\mu, D_v]$ which is proportional to the Einstein equations. (Interestingly he uses commuting spinors instead of anticommuting spinors).

– *F. Quevedo:*

Can the techniques you developed for the $2D$ superspace be used for computing for instance multiloop amplitudes for the string?

– *P. van Nieuwenhuizen:*

Yes since we solved the torsion constraints, the theory is ready to be used for any quantum computation. It would be too complicated to quantize a system with constraints.

– *M. McGuigan:*

If one had N=8 supergravity in two dimensions would its sigma model be over a manifold with octionic structure?

– *P. van Nieuwenhuizen:*

That is a hot issue. Nobody knows the answer to this question. I think there are many more models, but these models will not start with the scalar curvature but the curvature multiplied by some scalar field. You cannot do the Weyl rescaling in two dimensions, that is the one dimension where you cannot do it. So there is a new class of models. These models may not have a global limit in some sense so they do not have a corresponding super–algebra, and that would explain why all the papers by people who analyse the algebras tell you there is nothing beyond N=4. There are such examples in supergravity known, there is a model by Freedman and Schwartz for N=4 which does not have a rigid algebra. So there are probably a lot of new models which will soon appear. One way to discover these models is to take an N=4 or N=8 model in four dimensions and dimensionally reduce to two dimensions. You must get at least N=8 or N=16. So I think there are many more models. Whether they are useful is doubtful and I have nothing to say on that.

– *M. McGuigan:*

If a string is in some dimension outside the critical one, so that conformal invariance is broken by the anomaly, would there still be massless particles among its spectrum?

– *P. van Nieuwenhuizen:*

Below the critical dimension it is very difficult to prove or disprove that these models are consistent. Neveu, Gervais and others have done a lot of work on this problem. One will have an extra mode, the scalar Liouville mode which comes from gravity and will have to be kept in the theory. These particles will be massless because of supersymmetry. You can find in the papers of Fradkin and Tseytlin claims that Polyakov was wrong because the Liouville action should be massless. The mass term which Polyakov found should cancel and that claim is correct because in supersymmetry there are an equal number of bosons and fermions. (Actually, I believe that is not sufficient. The mass term cancels because the coefficients for the mass terms of the dilaton add up to zero).

– *A. Shapere:*

If a non–SUSY field theory of quantum gravity with all the nice properties of supergravity were to be found would this be a set back for the program of quantum gravity?

– *P. van Nieuwenhuizen:*

You mean a setback for supergravity? Definitely. Before I started working on supergravity I followed your line of thought. I looked at conformal gravity as possible theories of gravity. I was looking for other symmetries and a finite theory of gravity. It seemed that any symmetry might help to cancel infinities and then in 1976 we tried supergravity and that seemed to work at the one and two loop level and today nobody knows if at the three loop level it is divergent. The thing I like about SUSY is that in this Fermi–Bose symmetry if you take the commutator it leads to a translation. It is intrinsically a symmetry of spacetime structure. It is not a Yang–Mills symmetry which always acts at the same point. Thus we got a spacetime symmetry which indeed did cancel some (all?) quantum infinities. There is no other symmetry known which can do that. Whether the local Fermi–Bose symmetry must be used as in supergravity or in superstring theory, or in another way, remains to be seen.

– *A. Shapere:*

How about the field theory derived from $O(16) \times O(16)$ string theory which is not supersymmetric?

– *P. van Nieuwenhuilzen:*

If that theory was really acceptable then you would be right supersymmetry would be really superflous. However that theory is unstable.

– *P. Shotton:*

You mentioned that some theoreticians work with theories with negative dimensions. How do you interpret the concept of negative dimensions?

– *P. van Nieuwenhuizen:*

Parisi and Nicola Sourlas in Paris and also other people have thought about negative dimensions, and then they thought in the sense of supersymmetry that you could make sense of negative dimensions. Also in some Statistical Mechanics models you can sometimes introduce the notion of negative dimensions. I have asked people about it because I got these results in N=4 D=2 supergravity and I was not so happy with it. I have also wondered if you could get more complicated models in which the critical dimension changes, so that it goes up. However if one does these computations for non–linear sigma models, the critical dimension becomes always lower. This is well known from operator methods, but you get it also from the path integral. No one has been able to lift the critical dimension. At present, the concept of negative dimension in this context is nonsense to me, but some day someone may find a way to intepret it.

– *P. Shotton:*

You mentioned an anti–Majorana particle. What is that?

– *P. van Nieuwenhuizen:*

An anti–Majorana spinor is i times a Majorana spinor. It is a trivial remark. If you have a bosonic parameter, like the general coordinate parameter, you replace it by the corresponding ghost times the anti–commuting parameter; by convention we always take the ghost out the anti–ghost real. If you go to supersymmetry, then the supersymmetry parameter is replaced by the supersymmetric ghost field times Λ. If ξ_μ is real, then $\xi^\mu \rightarrow C^\mu\Lambda$. If C^μ is real, then since Λ and C^μ anti–commute, Λ must be pure imaginary. If you do the same for supersymmetry parameter $\epsilon^q = S^q\Lambda$, then if ϵ^a is real or Majorana and as Λ is pure imaginary we have that the ghost must be purely imaginary or (in a general representation) an anti–Majorana spinor.

– *J. Quakenbush:*

Why are BRST transformations global transformations?

– *P. van Nieuwenhuizen:*

Actually these are rigid transformations. Rigid means that the parameters at different points cannot be chosen freely. For example constants. They can also be space–time dependent but if the parameter is chosen at one point it is fixed at all other points. Global usually has a topological sense. BRS symmetry is a rigid symmetry because after you do gauge fixing you have lost the local symmetry. But after you fix the gauge and add ghost terms the remarkable thing is that you get a symmetry back with the parameters replaced by fields times a constant anticommuting parameter. As the parameter is a constant this is a rigid transformation. I did once write a paper on local BRS transformations, but I have found no use for it yet (see Nuclear Physics).

– *S. Carlip:*

A comment concerning the ambiguity of the functional integral measure. You assume that the measure should be invariant under coordinate transformations. This seems sensible. But De Witt has shown that this is not necessarily right. In theories in which the commutator of two transformations gives you a structure functional rather than a structure constant, you can start with the theory on the space of orbits and run the Fadeev–Popov procedure backwards to see what you get. You end up with a measure which is not invariant, but whose variation is proportional to the trace of the structure function. If you choose the wrong measure you get an ambiguity if you choose the right one, there is still an ambiguity because field redefinition changes the structure functions, but it is cancelled by a corresponding change in the Fadeev–Popov determinat. This may be relevant to the ambiguity you find.

– *P. van Nieuwenhuizen:*

The problem of a measure in function space is difficult and not yet solved. The naive measure discussed in the lecture seems to work and gives the right anomalies, but at a deeper level I do not really know why the question of measure is still unsolved. I know about De Witt's work. It appeared in the proceedings of "Les Houches" a few years ago.

– *M. Quiros:*

Is there any super–gauge where superghosts are decoupled?

– *P. van Nieuwenhuizen:*

That is a good question. I don't know the answer. At first thought it seems not possible because for vectors you go from the Lorentz to the axial gauge by dropping a derivative, whereas for the gravitino the usual gauge fixing term $\gamma \cdot \psi$ is already without derivative, so you cannot peel off another derivative.

– *M. Quiros:*

Concerning a previous question I believe that the only usefulness of SUSY at low energy is to solve the hierarchy problem, since the vanishing of the cosmological constant is uncorrelated to local SUSY. I do not see any other trouble to the $O(16) \times O(16)$ string theory.

– *P. van Nieuwenhuizen:*

This is a point of view one could have, that the hierarchy problem is the only thing. The Fermi–Bose symmetry that supergravity and superstring theory possess seems important for the finiteness of the quantum gravity and this should be kept in mind also. In the field theory limit these models I just mentioned with infinite sets of curvatures start looking very much like string theories. Perhaps they are finite and it deserves investigation.

– *A. Pasquinucci:*

Supergravity theories are thought to be non–renormalizable. Do you think that could exist some term like Chern–Simons terms in 1st order formalism which can cure that non–renormalizability?

– *P. van Nieuwenhuizen:*

People now understand why one and two loop calculations in supergravity where finite, because there is no possible counter–term. At three loops the situation is quite different and there is a possible counter–term and if it was forbidden by the presence of some extra symmetry, one would have a proof of finiteness, at least at three loop level. It is possible by adding extra terms to the action, such as Chern–Simons terms, one obtains new symmetries which cancel further divergences.

– *K. Meissner:*

Are the x–space and superspace approaches equivalent, I mean, is it only the question of appropriate constraints?

– *P. van Nieuwenhuizen:*

These approaches are exactly the same but only if you have auxiliary fields in the x–space there is a procedure called gauge completion which maps from x–space to superspace. For higher dimensions and extended supergravity one can show under certain mild conditions that there are no auxiliary fields. One must have all the auxiliary fields for x–space theory in order to have a superspace formalism. So it appears that there is no superspace formalism for x–space theories which have no auxilary fields. I will put a discussion of gauge completion in the lecture notes.

– *S. Giddings:*

I am told that in D=10 supergravity there is no way to write terms in the action that prevent 10 dimensional Minkowski space from being a solution to the equations of motion. In particular this fact appears relevant to superstrings. Can you explain the reason for this fact? In particular I am interested in terms that force compactification.

– *P. van Nieuwenhuizen:*

I know of no proof that it is not possible to introduce terms that prevent Minkowski space from being a solution. On the other hand, without a term like the cosmological constant it seems unlikely. Typically, in the terms, one may consider to add contain curvatures, and in the equations of motion one has terms containing curvatures R which vanishes for Minkowski space.

– *S. Giddings:*

You mentioned that it is not possible to integrate out the propagating fields in a theory. In string theory, one thinks of integrating out the massive fields to get an effective theory. At what point do problems arise with this procedure?

– *P. van Nieuwenhuizen:*

If one has massive fields Φ coupled with massless fields φ, one can integrate out Φ, thus obtaining a determinant of the form $(\Box + M^2)^{-1}$ expanding about M^2, one finds a nonlocal theory (due to arbitrarily many powers of $\Box$). It may be that any finite approximation is never a good approximation of the full theory.

– *G. Miele:*

Why is it not possible to build N=1 supergravity in more than eleven dimensions?

– *P. van Nieuwenhuizen:*

The usual argument goes as follows. If you reduce it to four dimensions then a spinor which has 32 components in eleven dimensions becomes eight spinors which have four components in four dimensions. Now, you can write down the representations of the superalgebra in 4 dimensions. For any N it follows from the theory of induced representations of a particular superalgebra. We find that in the case of N=8 model the multiplets are a kind of a cigar shape multiplets which run from helicity +2 to -2. If you would now go to higher dimensions eg. 12, 13, 14, and reduce to 4 dimensions, you would get an N=8 models and you look up these representations and again you get this cigar shape multiplet. But now they stick a little bit out of helicity +2 and -2. You get particles with spin larger than 2. It is generally believed that with a finite number of particles you cannot write down a consistent theory of particles with spin large than 2. These fields have to be massless, therefore they have to be gauge fields but very general arguments tell you that beyond energy momentum conservation. There is no other conservation law you can think of which would make these theories with higher spin consistent. Maybe there does not exist a global change but so far the models people have checked do not support it. There is a recent paper of Thierry–Mieg who claims that in 26 or 18 dimensions there are nice new string theories. These theories have a supersymmetric spectrum, but reduced to four dimensions they will give something like N=2000 with tremendously high spins. Thus I believe they do not exist as Lagrangians field theories.

FOUR-DIMENSIONAL SUPERGRAVITIES FROM SUPERSTRINGS

S. Ferrara

CERN-Geneva Switzerland
and
Physics Department, University of California, Los Angeles

ABSTRACT

Supergravity theories emerging as the point-field limit of various superstring compactifications are considered, and the higher order corrections to the standard supergravity Lagrangians are discussed. The structure of the effective Lagrangians for the recently constructed four-dimensional superstring models is also reported.

I. INTRODUCTION

The 1984 discovery of Green and Schwarz[1] of chiral superstring theories with anomaly free Yang-Mills groups has opened the way for the search of a theory encompassing the fundamental interactions of nature, including a consistent theory of quantum gravity. Intensive study of superstring theories over the last years[2] has revealed a large variety of possible "superstring vacua," at least in the framework of string perturbation theory. The number of possible consistent superstring models has increased enormously since the original work of Candelas, Horowitz, Strominger, and Witten,[3] based on Calabi-Yau compactifications. String compactifications on orbifolds[4] and more general constructions[5-9] of (heterotic) superstring in 10-d dimensions with Yang-Mills groups of rank 16+d have revealed a large variety of possible superstring models, which, with the present knowledge, seem all to be consistent, at least in a pertubative sense.

All these superstring theories, in the point-field limit, reduce to the Einstein theory of gravitation, coupled to ordinary matter fields of spin 0, 1/2, and 1. These are the string states which correspond to the "massless" sector of the theory. Integration over the massive stringy modes produces higher derivative modifications to the Einstein theory coupled to matter, with higher order terms in the gravitational and Yang-Mills curvatures.

Lectures given at the 24th International School of Subnuclear Physics, Erice, Sicily (1986).

The massless sector contains extra massless spin 3/2 excitations in the case of superstrings with some unbroken supersymmetry. However, there is a limit in the number N of these massless spin 3/2 excitations, depending on the particular superstring theory one is considering. In heterotic[10] superstrings, N can be at most 4 while, in type II superstring, N can be at most 8[2]. Here N will always refer to the number of unbroken supersymmetries in four-dimensions.

The additional request of having chiral fermions in four-dimensions, as well as "stability properties" of supersymmetric vacua against quantum effects, favor N=1 theories to be the most interesting ones for an extension of the "standard model" with a natural solution of the hierarchy problem[11]. The latter issue is related to the stability of scalar Higgs masses versus quantum effects at energies much below the Planck scale.

For theories in which one or more supersymmetry survives at low energies, superstring theories reduce to supergravity theories coupled to a certain number of matter supermultiplets. These theories will look very different depending on the special string compactifications. However, some general properties of the supergravity Lagrangians can be derived by use of some general symmetries of strings as well as the knowledge of the massless spectrum of each particular model.

In section II we will review the general structure of four-dimensional supergravity Lagrangians obtained by dimensional reduction of ten-dimensional superstrings.

In section III the string symmetries will be used to discuss the general forms of the Kahler potential and of higher order corrections. In section IV the supergravity Lagrangians obtained as point-field limits of four-dimensional superstrings will be described.

II. DIMENSIONAL REDUCTION OF TEN-DIMENSIONAL SUPERSTRINGS

A general way for approximating the effective D=4 superstring action which describes the interaction of the massless modes has been proposed by Witten[13]. More specifically, the strategy consists in a dimensional reduction of the 10D, N=1 supergravity Lagrangian and then performing a consistent truncation[12] by keeping only singlets with respect to some suitable group, isomorphic to a subgroup of the rotation group of the internal coordinates.

The simplest example was worked out in ref. (12), and it is just the extreme case of this procedure, obtained by keeping only singlets of $SU(3)_D = SU(3)_1 \otimes SU(3)_2$, in which $SU(3)_1$ is a subgroup of SO(6), the rotation group of the internal coordinates X_I and $SU(3)_2$ is in G where $G=E_8 \times E_8'$ or SO(32).

In the present section we generalize this approach in order to obtain more general D=4 effective Lagrangians for the massless modes preserving some space-time supersymmetry[14]. In particular, with gauge group $E_8 \times E_8'$ one can obtain effective N=1 Lagrangians with several families in the (27) irreducible representation of $E_6 \subset E_8$.

The one-family model reproduces the original example of ref. (12). The maximal model with nine-families corresponds to the effective Lagrangian for the untwisted sector of the T_6/Z_3 orbifold where T_6 is the six-dimensional torus and Z_3 is the center of $SU(3)^4$. Lagrangians with an intermediate number of families can be obtained as well and can be derived as truncation of the maximal model.

A common property of these Lagrangians is their no-scale structure[15], namely, the (semi)-positivity of the scalar potential and its flatness along some (complex) directions of the scalar field variables.

The N=1 Lagrangians are all obtained by truncation of a N=4 Lagrangian[16] with gauge group $U(1)^6 \times E_8 \times E_8'$ which corresponds to trivial torus compactification. There is an extra $U(1)^6$ gauge symmetry coming from the supergravity sector which is broken to U(1) in N=2 compactifications or completely broken in N=1 compactifications. These extra U(1) symmetries are associated with the "graviphotons," i.e., the spin 1 partners of the graviton in the N-extended graviton multiplet.

In the sequel we will only consider the bosonic part of the Lagrangian. In fact in all supersymmetric theories the fermionic part is uniquely fixed by supersymmetry once the bosonic sector is known.

In general we may split the 10D indices $\hat{\mu} = \mu, I$ ($\mu=1,\ldots,4$, $I=1,\ldots,6$) so the D=10, N=1 bosonic supergravity fields are[17]

$$g_{\mu\nu},\ g_{\mu I},\ g_{IJ},\ B_{\mu\nu},\ B_{\mu I},\ B_{IJ},\ \phi\quad . \tag{2.1}$$

The D=10 gauge non-singlet bosonic fields are

$$A^{\alpha}_{\mu},\ A^{\alpha}_{I} \tag{2.2}$$

We confine our discussion here to N=1, D=10 supergravity, coupled to Yang-Mills matter, although the same strategy may be used for N=2, D=10 chiral and non-chiral supergravities. The latter case would correspond to some compactification of type II strings in the point-field limit[18]. In equation (2.1) $g_{\mu\nu}$, $B_{\mu\nu}$, and ϕ are respectively the metric tensor, the antisymmetric tensor, and the dilaton. Since these fields are SO(6) singlets they will survive to any reduction of the type considered here. Compactifications in which $b_{\mu\nu}$ and ϕ become massive because of one-loop effects, while preserving N=1 supersymmetry, have been recently considered in the context of SO(32) superstrings[19].

The simplest dimensional reduction corresponds to a T^6 torus compactification. This demands that we keep all modes given by equations (2.1) (2.2), and it results in an N=4, D=4 supergravity theory coupled to 502 N=4 (vector) matter supermultiplets with gauge symmetry $U(1)^6 \times E_8 \times E_8'$, a group of rank 22. More general groups of rank 22 will appear in the case of four-dimensional superstrings corresponding to toroidal stringy compactifications, first considered by Narain[5].

The N=4 supergravity theory obtained from the dimensional reduction of 10D-supergravity has a scalar sector whose derivative interactions describe a σ-model based on the coset space:

$$\frac{SU(1,1)}{U(1)} \times \frac{SO(6,6+n)}{SO(6) \times SO(6+n)} \quad \text{with } n = \dim G = 496 \tag{2.3}$$

For n=0 we obtain the coset space-structure of the pure supergravity sector first considered by Chamseddine[17]. The SU(1,1)/U(1) part, common to all compactifications, is related to the $B_{\mu\nu}$ and ϕ massless modes. To prove equation (2.3), we first write the 10D Lagrangian, neglecting the contribution of the 508 vector fields A^{α}_{μ}, $g_{\mu I}$, $B_{\mu I}$. After a rescaling of the D=4 metric $g_{\mu\nu}=\Delta^{1/4}\, g_{\mu\nu}{}^4$ with $\Delta = \det g_{IJ}$, we obtain

$$\mathcal{L}(\sqrt{g})^{-1} = -\frac{1}{2} R - (\partial_\mu\phi)^2 - \frac{1}{16} (\partial_\mu \ell g\Delta)^2 + \frac{1}{8}\, \partial_\mu g_{IJ} \partial_\mu g^{IJ}$$

$$- \frac{3}{4} e^{2\Phi} H^2_{\mu IJ} - \frac{3}{4} \Delta e^{2\Phi} H_{\mu\nu} - \frac{1}{4} e^{\Phi} F^2_{\mu I}$$

$$- \frac{1}{4} e^{\Phi} \Delta^{-1/2} F^2_{IJ} - \frac{3}{4} e^{2\Phi} \Delta^{-1/2} H^2_{IJK} \quad . \tag{2.4}$$

The last two terms in equation (2.4) give rise to the N=4 scalar potential which is manifestly positive semidefinite. To see the structure of the nonlinear σ-model we make the following redefinitions[14]:

$$S = \sqrt{\Delta e^{\Phi}} + 3i\sqrt{2}D \quad . \tag{2.5}$$

When D is the dual of the $B_{\mu\nu}$ field $[e^{2\Phi}\Delta H_{\mu\nu\rho} = \epsilon_{\mu\nu\rho\sigma}\partial^{\sigma}D]$,

$$T = T_{IJ} = e^{-\Phi} g_{IJ} + A^{\alpha}_{I} A_{\alpha_J} + \sqrt{2} B_{IJ} \quad . \tag{2.6}$$

The complex coordinate S parametrize the SU(1,1)/U(1) Kahler manifold, based on the Kahler potential,

$$J_s = -\log\,(S+S^*) \tag{2.7}$$

The real coordinates T_{IJ}, A^{α}_{I} parametrized the manifold $\frac{SO(6,+6+n)}{SO(6) \times SO(6+n)}$.

The geometric structure becomes manifest if one uses projective variables $P_{IA} = (P^1_{IJ}, P^2_{I\alpha})$ defined by

$$T_{IJ} = \frac{1}{2}\left[(1 + P^1)(1 - P^1)^{-1}\right]_{IJ}$$

$$A_{I\alpha} = \left[(1 - P^1)^{-1} P^2\right]_{I\alpha} \quad , \tag{2.8}$$

so that the kinetic terms reduce to

$$- \frac{\partial_{\mu}S\partial_{\mu}S^*}{(S+S^x)^2} - \frac{1}{2}\,\mathrm{Tr}\,(1-P^TP)^{-1}\,\partial_{\mu}P(1-P^TP)\partial_{\mu}P^T \; . \tag{2.9}$$

We will consider now consistent truncations of the D=10 theory which reduces the number of supersymmetries. The spin 3/2 gravitinos transform as 4 of SU(4) ≃ SO(6). So, any subgroup of SU(4) under which some gravitinos are singlets correspond to a truncation and in some cases to a true compactification. The number of singlet gravitinos correspond to the number of unbroken supersymmetries.

As an intermediate step we may obtain an N=2 supergravity by decomposing SU(4) → SU(2)' × SU(2) and by keeping only SU(2)' singlets. A weaker requirement, which corresponds to an orbifold compactification on $T_4/Z_2 \times Z_2$, is obtained by retaining only Z_2 singlets under the center of SU(2)'. In this case the unbroken gauge group is $U(1)^3 \times SU(2) \times E_7 \times E_8'$ with an additional U(1) gauge symmetry coming from the graviphoton.

The extended supergravity theories so far considered are trivially anomaly-free because of the number of surviving supersymmetries. However, as we will see now, in the N=1 case, families are in complex representations of the gauge groups G so the resulting massless sector may show up gauge anomalies.

In order to obtain the maximal model with N=1 supersymmetry we pick up the smallest group under which there is only one singlet gravitino. This is the center Z_3 of SU(3). The Z_3 singlet scalars coming from the supergravity sector are

$$g_{i\bar{j}},\; B_{i\bar{j}},\; \Phi,\; D \tag{2.10}$$

where we used a complex notation for the internal coordinates corresponding to the decomposition $6 \to 3 + \bar{3}$ of SU(4) into SU(3). The charged fields decompose under SU(3) × E_6 as follows

$$A_I^\alpha \to C_3^{(\bar{3},\, \overline{27})},\; C_3^{(3,\, 27)},\; C_3^{(1,\, 78)},\; C_3^{(8,\, 1)} \,. \tag{2.11}$$

The diagonal Z_3 singlets are the complex fields

$$C_i^{(m,\, a)} \text{ with } a\epsilon 27 \; i\epsilon\bar{3} \text{ of } SU(3')',\; m\epsilon 3 \text{ of } SU(3) \,. \tag{2.12}$$

The surviving gauge fields are in the adjoint of SU(3) × E_6 × E_8'. The present Yang-Mills sector contains nine (27) families which are triplets under the horizontal gauge symmetry SU(3).

The gravity sector corresponds to 10 chiral multiplets which are singlets under the surviving gauge group. We observe at this point that the present set of massless states has SU(3) gauge anomalies. Note that in Calabi-Yau compactifications the SU(3) gauge group is completely broken,[3] so no anomaly problem arises. The simplest way to cancel the SU(3) gauge anomalies is to have 81 additional chiral multiplets in the $\bar{3}$ of SU(3). These additional chiral fermions come from the "twisted" sector[4] of T_6/Z_3 stringy orbifold compactification and cannot be obtained as dimensional reduction of the 10D point-field Lagrangian. The twisted sector contains 27 copies of $(\bar{3},1) + (1,27)$ massless fermions.

The N=1 supergravity Lagrangian describing the effective interactions of the massless modes is entirely specified in terms of the Kahler potential $J(Z,Z^*)$, the superpotential $g(Z)$ (which is an analytic function of the complex scalar fields Z), and the Yang-Mills metric $f_{AB}(Z)$[20]. From the D=10 reduction of the supergravity Lagrangian we can only determine these functions for the untwisted sector alone. In order to determine the Kahler potential we define complex fields $T_{i\bar{j}}$ and S:

$$T_{i\bar{j}} = g_{i\bar{j}}e^{-\phi} - C_i^{ma}C_{\bar{j}}^{*\bar{m}\bar{a}} - i\sqrt{2}B_{i\bar{j}} \quad , \tag{2.13}$$

$$S = \det g_{i\bar{j}}e^{\phi} + 3i\sqrt{2}D \quad . \tag{2.14}$$

Then the N=1 Kahler potential is $(Z{=}S,\ T_{i\bar{j}},\ C_j^{ma})$

$$J(Z,Z^x) = -\ell og(S+S^x) - \log\det(T_{i\bar{j}} + T_{i\bar{j}}^* - 2C_i^{ma}C_{\bar{j}}^{*\bar{m}\bar{a}}) \,, \tag{2.15}$$

while the superpotential is[14]

$$g(Z) = d_{abc}\,\epsilon_{mnl}\,\epsilon^{ijk}\,C_i^{ma}\,C_j^{nb}\,C_k^{\ell c} \quad . \tag{2.16}$$

The gauge kinetic function is $f_{AB} = \delta_{AB}S$.

The Kahler metric obtained from (2.15) corresponds to the Kahler manifold[14]

$$\frac{SU(1,1)}{U(1)} \times \frac{SU(3,\ 3+n)}{SU(3) \times SU(3+3n) \times U(1)} \quad \text{with } n=27 \,. \tag{2.17}$$

Obviously this is a truncation of the N=4 manifold given by (2.3). We observe that the geometry of this Kahler manifold guarantees the positiveness of the potential[21] for arbitrary superpotentials of the form $g_1(S) + g_2(C)$. This allows the mechanism of gauging condensation. If we retain only SU(3) singlets this would kill eight of the nine $T_{i\bar{j}}$ chiral multiplets and eight of the nine chiral families. It would also eliminate the SU(3) gauge vector multiplets. The resulting Lagrangian would be the Lagrangian obtained by Witten[12], based on a scalar manifold of the type

$$\frac{SU(1,1)}{U(1)} \times \frac{SU(n,1)}{SU(n) \times U(1)} \quad n=27 \ . \tag{2.18}$$

This Lagrangian is the approximated low-energy Lagrangian for a Calabi-Yau manifold with one chiral family.

III. STRING SYMMETRIES AND HIGHER ORDER CORRECTIONS

In this section we describe general restrictions on the superstring effective Lagrangian coming from string symmetries and local supersymmetry. These restrictions do not actually depend on any particular compactification scheme, so they may be equally valid for four-dimensional compactifications of higher dimensional superstrings as well as genuine four-dimensional strings.

We divide the massless chiral multiplets of general N=1 four-dimensional superstring in three basic sets, one associated with dilation, called S field. Another set associated with possible breathing modes of the internal manifolds, devoted by T and a set of charged chiral fields C_i transforming under some representation of the four-dimensional Yang-Mills symmetry G. These sets of fields have been already introduced in the previous section in some particular compactification schemes.

The classical string effective action has a global scale invariance[12] (covariance) under a shift of the metric and of the S chiral field,

$$g_{\mu\nu} \rightarrow \lambda g_{\mu\nu}, \ S \rightarrow \lambda S, \ \mathcal{L} \rightarrow \lambda \ \mathcal{L} \ , \tag{3.1}$$

the T and C multiplets, as well as the gauge vector multiplets, V being inert under this scale transformation.

At the n-string loop the Lagrangian scales as

$$\mathcal{L}_{n\text{-loop}} \rightarrow \lambda^{1-n} \ \mathcal{L}_{n\text{-loop}} \tag{3.2}$$

under the same transformation.

There is another scale symmetry associated to the dilation mode of the internal metric T of the manifold. The classical[12,22] Lagrangian scales as

$$\mathcal{L} \rightarrow r^{-1/2} \mathcal{L} \tag{3.3}$$

under the following transformation of the S, T, and C fields:

$$S \rightarrow r^{-1/2} S, \ T \rightarrow r^{+1/2} T, \ C^i \rightarrow r^{+1/4} C^i; \tag{3.4}$$

the n-string loop Lagrangian scale as[23]

$$\mathcal{L}_{n\text{-loop}} \rightarrow r^{n-1/2} \ \mathcal{L}_{n\text{-loop}} \tag{3.5}$$

under the same scale transformation.

These symmetries are meant to be valid for the full superstring effective action, which means that also higher derivative couplings will have to respect these symmetries.

In this section we will show that these symmetries take a particulary simple form if one uses a superconformal formulation[24] for the standard N=1 supergravity Lagrangian as well as for higher derivative supergravity couplings[25].

Let us introduce, for convenience of the reader, some superconformal preliminaries[30,24].

In the superconformal tensor calculus (or superspace) formulation of matter coupling to N=1 supergravity, superfields have definite conformal properties.

In particular the standard supergravity formulation corresponds to the so-called "minimal formulation"[26]. This uses a compensator multiplet S_0 whose last component $u = S - iP$ corresponds to the scalar auxiliary fields of minimal supergravity. Each conformal superfield has definite Weyl and chiral weight. In particular chiral multiplets have equal chiral and Weyl weight while real vector multiplets have vanishing chiral weight.

The gravitational super derivative $D\alpha$ has chiral and Weyl weight equal to one-half, while the D and F superconformal densities have respectively Weyl weight two and three. Physical chiral multiplets are taken to have vanishing Weyl weight while the compensator S_0 has Weyl weight one.

Also the so-called real "linear" multiplet[27] L, containing an antisymmetric tensor field, has Weyl weight two.

In theories with a Green-Schwarz mechanism the linear multiplet L is supported to transform non-trivially under gauge gravitational and Yang-Mills transformations.

This implies the introduction of a "covariant" multiplet[28, 29]

$$\hat{L} = L - \Omega_3 \tag{3.6}$$

where Ω_3 is called a Chern-Simons supermultiplet, satsifying the Bianchi-identity[29]

$$\Sigma(\Omega_3) = TrW^2 \tag{3.7}$$

where W^2 is the square of the Yang-Mills or Weyl field strength multiplets W_α and $W_{\alpha\beta\gamma}$[30]. $\Sigma(\bar{\Sigma})$ denotes the chiral (antichiral) projection of a general multiplet, so that by definition,

$$\Sigma(L) = \bar{\Sigma}(L) = 0 \tag{3.8}$$

which implies

$$\Sigma(\hat{L}) = -TrW^2. \tag{3.9}$$

We note that because of equation (3.7) the Chern-Simons form satisfies the property

$$\int d^8\Omega\ (S + \bar{S})\Omega_3 = Re\int d^6\Omega\ STrW^2 \tag{3.10}$$

($d^8\Omega$ and $d^6\Omega$ denote full and chiral superspace measures), where S is any chiral multiplet. Equation (3.10) allows one to convert a linear multiplet with Green-Schwarz mechanism into a chiral multiplet (S) coupled to gauge-fields in a manner dictated by equation (3.10).

As a simple exercise we rewrite the simplest N=1 supergravity Lagrangian obtained from superstring dimensional reduction as discussed in the previous section.

In the case of a truncation which retains only SU(3) singlets, the form of the Kahler potential is[12]

$$J = -\ell og(S + \bar{S}) - 3\ell og\ (T + \bar{T} - C_i C^{*i}) \qquad (3.11)$$

while the superpotential term is

$$g(C) = d_{ijk}C^i C^j C^k$$

and d_{ijk} is the symmetric coupling of the 27-dimensional representation of E_6. The standard superspace expression of the full low-energy Lagrangian is

$$-\int d^8\Omega\phi So\bar{S}o + \text{Re}\int d^6\Omega S\text{Tr}W^2 + \text{Re}\int d^6\Omega g(C)S_0^3 \qquad (3.12)$$

where

$$\phi = (S + \bar{S})^{1/3}(T + \bar{T} - C_i\bar{C}^i) = (S + \bar{S})^{1/3}\tilde{\phi} \qquad (3.13)$$

and superconformal geometry has been used.

Let us discuss the symmetries of this superspace Lagrangian. Under the λ symmetry [equation (3.1)] the compensator S_0 must transform as

$$So \to \lambda^{1/3}So \qquad (3.14)$$

while under the r symmetry the compensator must transform as

$$So \to So\ r^{-5/12}. \qquad (3.15)$$

We see that the only freedom left from the λ and r symmetries is a modification to the Kahler part of the form[22]

$$\phi \to \phi f\left(\frac{C}{\sqrt{T+T^*}}, \frac{C^*}{\sqrt{T+T^*}}\right) \qquad (3.16)$$

when we have also used the axion symmetries

$$T \to T + ic \quad \text{and} \quad S \to S + ic. \qquad (3.17)$$

The axion symmetry on the S field is needed in order to convert it into a linear multiplet L. In fact if S appears in the Kahler potential only through a $S + \bar{S}$ dependence, we can rewrite a Lagrangian equivalent to (3.12) as follows[27]:

$$-\int d^8\Omega\ U^{1/3}\ \tilde{\phi}\ So\bar{S}o + \int d^6\Omega U\Omega_3 + \text{Re}\int d^6\Omega g(C)S_0^3 - \int d^8\Omega LU\ , \qquad (3.18)$$

where L is a real number multiplet defined by equation (3.8) and U is a general real superfield. Variation of (3.18) with respect to L gives

$$U = S + \bar{S} \qquad (3.19)$$

and leads back to equation (3.12). If we vary instead with respect to U, then we get the duality transformed Lagrangian. This superspace Lagrangian, up to a normalization factor, is

$$\int d^8\Omega \left(\frac{\hat{L}}{So\bar{S}o\tilde{\phi}}\right)^{-1/2} \tilde{\phi}\ So\bar{S}o + \text{Re}\int d^6\Omega g(C)S_0^3. \qquad (3.20)$$

When $\hat{L} = L - \Omega_3$ and $\tilde{\phi} = T + \bar{T} - Ce^V\bar{C}$. The first term in equation (3.20) can be immediately generalized to n-loops as follows

$$\int d^8\Omega \left(\frac{\hat{L}}{So\bar{So}\tilde{\phi}}\right)^{(3/2)n-1/2}(So\bar{So}\tilde{\phi})\tilde{\phi}^n\ f_n\left(\frac{C}{\sqrt{T+T^*}},\frac{e^V C^*}{\sqrt{T+\bar{T}}}\right) \tag{3.21}$$

while the superpotential term is unrenormalized[22]. There is an extra one-loop term compatible with both λ and r symmetries, namely,[23]

$$\mathrm{Re}\int d^6\Omega\ (\alpha T + \beta C^2)W^2 \tag{3.22}$$

In absence of the r symmetry, associated with the T field, equations (3.21) and (3.22) take the form

$$\int d^8\Omega \left(\frac{\hat{L}}{So\bar{So}}\right)^{3/2n-1/2}(So\bar{So})\ f_n\ (C_i,(e^V C^*)^i) \tag{3.23}$$

$$\mathrm{Re}\int d^6\Omega f(C_i)\mathrm{Tr}W^2 \tag{3.24}$$

where $f(C_i)$ is a gauge invariant chiral function constructed out of the C_i fields. At one-loop there is a possible extra contribution connected to anomaly cancelling terms[19,31,29]

$$cLV \tag{3.25}$$

where V is a U(1) gauge field with anomalous fermion representation.

So far we have confined our discussion to lowest derivative interactions, which modify the Kahler potential of the standard supergravity form, due to string loop effects.

One can equally discuss the general structure of higher derivative interactions coming from the integration of the massive stringy modes[25]. For simplicity, let us confine ourselves to fourth derivative interactions, although our reasoning can be extended to invariant terms with derivatives and curvatures of any order.

Quartic derivative interactions are already present in expressions (3.20) and (3.23) due to the presence of the Lorentz Chern-Simons form Ω_3^L in the definition of L. Other four-dimensional interactions can come from terms of the form

$$\int d^8\Omega\ \frac{W^2\bar{W}^2}{\hat{L}^3}\left(\frac{\hat{L}}{So\bar{So}\tilde{\phi}}\right)^{-1/2} So\bar{So}\tilde{\phi} \tag{3.26}$$

where W is in Yang-Mills field strength. We note that the dependence in equation (3.26) on the S_0 and $\tilde{\phi}$ fields is fixed by the λ and r symmetries while the $\hat{L}$ dependence is fixed by superconformal symmetry since the quartic term $W^2\bar{W}^2$ has Weyl weight six. The n-loop modification to the quartic invariant given by (3.26) is the same as equation (3.21) with the inclusion, in the intergrand, of the term

$$\frac{W^2\bar{W}^2}{\hat{L}^3} \tag{3.27}$$

which is inert under λ, r and superconformal symmetry.

It can be shown that higher curvature terms modify the standard supergravity potential; however, it is a general fact that supersymmetric extrema remain unaffected by higher curvature terms, at least in a perturbative sense[32].

IV. EFFECTIVE LAGRANGIANS FOR FOUR-DIMENSIONAL SUPERSTRING MODELS

In this latter section we will briefly report on the structure of effective Lagrangians for superstring theories directly formulated in four space-time dimensions.

It is well known that D=10, N=1 superstrings reduce, in the point-field limit, to D=10 supergravity coupled to Yang-Mills matter with gauge group G $E_8 \times E_8'$ or SO(32).

These gauge groups are actually the only possible gauge symmetries compatible[1] with anomaly freedom of D=10, N=1 supergravity. In the framework of superstrings, if one first takes the point-field limit, and then studies the compactification of the underlying field theory, it is unavoidable that the residual gauge group in D=4 dimension be a subgroup of $G \times U(1)^6 \times U(1)^6$. The twelve additional U(1) factors correspond to the abelian symmetries coming from possible massless vector modes associated with the graviton and the antisymmetric tensor field of D=10 superstrings.

In the case of a surviving N=1 supersymmetry in four-dimension, with the standard embedding of the space-connection into the gauge group G, the residual gauge group is a subgroup of $E_6 \times E_8 \times SU(3)$ or $SO(26) \times U(1) \times SU(3)$. The SU(3) gauge factor can be present in the case of orbifold compactifications[4] while it is broken for Calabi-Yau type of compactification[3].

All these compactifications, as far as the gauge group is concerned, can be studied in the point-field limit, i.e., in a regime where the Kaluza-Klein compactification radius is much smaller than the string characteristic scale $\alpha'^{1/2}$.

However, quite recently, starting with the work of Narain[5], it has been realized that stringy compactifications may exist in which the four-dimensional gauge group is a group of rank 22 which is larger than the gauge group obtained from trivial compactification of the point-field supergravity theory.

For instance in the N=4 case the gauge symmetry $SO(32) \times U(1)^6 \times U(1)^6$ may be changed to $SO(44) \times U(1)^6$. For N=4 theories the generic Yang-Mills group G is a product of simply laced groups of total rank 22. Other possible gauge groups[5] beyond SO(44) are for examples $E_8 \times SO(28)$, $E_8 \times E_6 \times E_7 \times SU(2)$, $E_8 \times E_6 \times E_6 \times SU(3)$, $E_8 \times E_6 \times E_6 \times SO(4)$.

All these N=4 theories have been shown[6] to be toroidal compactifications of the D=10 heterotic string[10] where some of the background fields take a constant v.e.v.

The low-energy limit of N=4 four-dimensional superstrings can be obtained by noticing the fact that the N=4 supergravity Lagrangian coupled to N=4 Yang-Mills matter multiplets is almost unique. The only freedom in the N=4 Lagrangian is the further gauging of the six graviphotons and some extra "phases" if the gauge group G is not simple[16]. However, for the case of toroidal compactifications, the graviphotons correspond to trivial U(1) factors and for a simple group SO(44) no relative gauge coupling constants or phases are possible. In this case the N=4 low-energy effective theory is completely specified.

The main difference of this theory, with the one obtained by trivial torus compactifications and studied in section III, is the absence[33] of the neutral vector multiplets whose scalar T fields correspond to the breathing modes of the internal manifold.

In the case of compactifications a la Narain[5] the radius of compactifications is fixed and no-dilatation symmetry related to the breathing modes is present. This is the r symmetry considered in section III for the case of the Witten dimensionally reduced Lagrangian. Said in another way the N=4 multiplets whose vector components were the $g_{\mu i}$ and $b_{\mu i}$ fields, have become "charged" and they now belong to a certain subgroup of the simple gauge group SO(44).

Let us discuss in some detail the symmetries of the N=4 supergravity theory with gauge group SO(44). The scalar fields of these theories parametrize the coset space

$$\frac{SO(6,\ n)}{SO(6)\ \times\ SO(n)} \times \frac{SU(1,1)}{U(1)} \tag{4.1}$$

where n=dimG and $\frac{SU(1,1)}{U(1)}$ is the coset space of the pure N=4 supergravity sector.

The non-compact symmetry is realized non-linearly on the scalar fields and as a duality transformation on the 6 + N vector fields. The symmetry is broken by the G gauge coupling terms. The only case where the non-compact symmetry is unbroken is for the gauge group $G = U(1)^{22}$. In this case the non-compact symmetry becomes SO(6,22) which is precisely the group associated with the Lorentzian lattices considered by Narain[5].

Recently the Narain construction has been extended to theories with residual N=2 and N=1 supersymmetries[7,8,9]. In these constructions the even-lattices considered before are generalized to odd Lorentzian "charged lattices". A simpler way of constructing four-dimensional string models is to introduce 2D-free fermions as internal degrees of freedom, carrying a non-linear realization of world-sheet supersymmetry[34].

Modular invariance and factorization strongly constrain the permitted choices of the fermionic spin structures. If we consider a N=4, 4D model, N=2 models can be constructed through a Z_2 projection[7,8] with the following charactistics:

(a) There is a sector of the N=2 model which is obtained by a Z_2 truncation of the corresponding N=4 model. We call this sector the "untwisted sector" (UTS) in analogy with Z_2 orbifolds.

(b) There is an additional sector, necessary for modular invariance, which cannot be obtained by Z_2 truncation of the N=4 model. This sector is by itself Z_2 symmetric and we call it "twisted sector" (TS). N=1 models can be obtained with a further Z_2' protection.

In some of the N=2 and N=1 models the TS sector contains only massive states, and therefore does not contribute to the effective low-energy theory. In those cases, the entire N=2 and N=1 effective theories are given once the N=4 theory is specified and the action of Z_2 and $Z_2 \times Z_2'$ defined.

However, for these theories no net number of chiral families survive. To get chiral families one must include a massless twisted sector. In this case the N=2 and N=1 theories cannot be obtained by a truncation of the N=4 theory. Nevertheless, in some cases, the N=1 theory can be obtained by a single Z_2 truncation from the N=2 theory once the interactions of twisted and untwisted N=2 states are known.

Let us examine the quantum numbers and properties of the massless twisted and untwisted states in the case of the "minimal model" based on the simple gauge group SO(44) for the N=4 theory. Similar considerations can be applied to all other models.

The massless states of the N=4 theory are the N=4 supergravity multiplet and n =dim G vector multiplets. The N=4 theory has a global SU(4) symmetry in which the SO(4) ≃ SO(6) indices can be regarded as "compactified" internal indices. The N=4 supergravity has fields[16]

$$e_{a\mu},\ \psi_{\mu i},\ A_{\mu ij},\ \chi^{i},\ S \qquad (i,j = 1,\dots,4) \tag{4.2}$$

and the gauge multiplets has fields[16]

$$V_{\mu}^{A},\ \lambda_{i}^{A},\ \phi_{ij}^{A},\ A=1,\dots \dim G. \tag{4.3}$$

The graviphoton and the matter scalar fields are in the six-dimensional representation of SO(6), while the fermions are in the 4-dimensional representation of SU(4).

To get a Z_2 and $Z_2 \times Z_2'$ truncation of the N=4 theory we have to show how the Z_2 projections act on the gauge group SO(44), as well as on the SU(4) indices. Let us pick up a Z_2 in SU(4) as follows

$$Z_2:\quad \alpha=\begin{pmatrix}1 & & & \\ & 1 & & \\ & & -1 & \\ & & & -1\end{pmatrix},\ 1 \qquad \alpha^2 = 1\ . \tag{4.4}$$

The Z_2 truncation of the N=4 theory, which retains only Z_2 singlets, will give a N=2 theory since there are only two gravitinos which are Z_2 singlets.

The action of Z_2 on the gauge group G is obtained as follows: we decompose SO(44) into $SO(n_A) \times SO(n_B)$ with $n_A + n_B = 44$. [Modular invariance also requires: $n_A = 2r + 8m$ (n_A, $n_B \neq 0$), $r \in (0,1,2)$, $m \in (0,1,2,3,4,5)$.] Then

$$n_V = (n_A,1) + (1,n_B). \tag{4.5}$$

Under the Z_2, α element $n_A \to n_A$, $n_B \to -n_B$, where n_V, n_A, n_B are the vector representations of SO(44), $SO(n_A)$, $SO(n_B)$ respectively. Equations (4.4) and (4.5) define the action of Z_2 on the AdjG and on the 6-dimensional representation of SU(4).

$$\alpha{:}\mathrm{AdjG} \to \mathrm{AdjG}_A + \mathrm{AdjG}_B + \alpha(n_A,\ n_B)$$

$$\alpha{:}I_6 \to i_2,\ \alpha i_4 \tag{4.6}$$

where i_2 = (0,1) (2,3) and i_4 = (0,2) (0,3) (1,2) (1,3). The scalar fields which are Z_2 inert are therefore[35]

$$(\mathrm{AdG}_A)_{i_2},\ (\mathrm{AdG}_B)_{i_2},\ (n_A,\ n_B)_{i_4},\ S. \tag{4.7}$$

The first two are the "complex" scalar partners of the vector fields of $SO(n_A) \times SO(n_B)$ in N=2 vector multiplet, while the third ones are hypermultiplets in the (vector, vector) representation.

The S field is the scalar partner of one combination of the $A_{\mu(0,1)}$, $A_{\mu(2,3)}$ vectors while the other combination is the N=2 graviphoton.

In the N=2 theory the "superpotential term" is cubic and of the form (YYV) where Y is the hypermultiplet and V is the vector multiplet. The only thing which remains to be determined in order to specify the theory is the Kahler manifold of the $\dim G_A + \dim G_B + 1$ vector multiplets and the quaternionic manifold of the $n_A n_B$ hypermultiplets.

These manifolds can be obtained by a truncation of the N=4 manifold (see: equation (4.1)) under the Z_2 action. Here we only report the result. The Kahler manifold is[35,36,37]

$$\frac{SU(1,1)}{U(1)} \times \frac{SO(2, \dim G_A + \dim G_B)}{SO(2) \times SO(\dim G_A + \dim G_B)} \tag{4.8}$$

while the quaternionic manifold is

$$\frac{SO(4, n_A n_B)}{SO(4) \times SO(n_A n_B)}. \tag{4.9}$$

We note that the Kahler manifold given by (4.8) is indeed compatible with N=2 local supersymmetry,[38] while the quaternionic manifold in (4.9) is one of the spaces considered by Wolf[39]. The only freedom of the N=2 theory is the relative strength of the two gauge couplings of the two factor groups $SO(n_A)$ and $SO(n_B)$.

To further truncate the theory to a N=1 supersymmetric Lagrangian we have to perform another Z_2 projection. For this purpose we introduce a second Z_2 embedded in SO(4) as follows[35]

$$Z_2: \quad \beta=\begin{pmatrix} 1 & -1 & & \\ & & 1 & -1 \end{pmatrix}, \quad 1 . \tag{4.10}$$

Now the $Z_2 \times Z_2'$ group has elements α, β, $\alpha\beta$, 1. The $Z_2 \times Z_2'$ projection breaks N=4 to N=1 supersymmetry since only one gravitino is surviving the second Z_2' projection.

To see how the $Z_2 \times Z_2'$ group acts on the gauge group we decompose SO(44) into four factors[8]

$$SO(44) \to \prod_{i=0}^{3} SO(n_i), \quad \sum_i n_i=44 \quad (n_0=2 + 4\kappa, \; n_0+n_i=0 \bmod 8) \tag{4.11}$$

so that

$$n_V = n_0 + n_1 + n_2 + n_3 \tag{4.12}$$

under $Z_2 \times Z_2'$ we have

$$\begin{aligned} n_0 &= n_0 \\ n_1 &= \alpha n_1 \\ n_2 &= \beta n_2 \\ n_3 &= \alpha\beta n_3 \quad . \end{aligned} \tag{4.13}$$

For $\alpha = 1$ we would get back to the N=2 theory.

Now under the action of $Z_2 \times Z_2'$ the adjoint and six-dimensional representations of SO(44) and SO(6) transform as follows

$$\begin{aligned} \mathrm{Adj}G \to & \sum_i \mathrm{adj}G_i + \alpha(n_2,n_3) + \alpha(n_0,n_1) \\ & + \beta(n_1,n_3) + \beta(n_0,n_2) + \alpha\beta(n_0,n_3) + \alpha\beta(n_1,n_2) \end{aligned} \tag{4.14}$$

$$I_6 \to \alpha i_1 + \beta i_2 + \alpha\beta i_3 \tag{4.15}$$

where $i_2 = (0,2)(1,3)$, $i_3 = (0,3)(1,2)$.

Therefore the $Z_2 \times Z_2'$ scalar singlets are the fields with label

$$\left[(n_2,n_3) + (n_0,n_1)\right]_{i_1} \tag{4.16}$$

from the N=2 vector multiplets and the fields[35]

$$\left[(n_0,n_3) + (n_1,n_2)\right]_{i_3}, \quad \left[(n_0,n_2) + (n_1,n_3)\right]_{i_2} . \tag{4.17}$$

from the hypermultiplets.

The three sets of chiral multiplets given by equations (4.16) and (4.17) exhaust the massless states of the untwisted sector of the N=1 superstring models based on gauge group $\prod_i SO(n_i)$.

To determine the effective Lagrangian one has to calculate the Kahler potential, the superpotential and the gauge kinetic function which define the N=1 4D, standard supergravity theory[20].

The Kahler manifold of the chiral multiplets can be obtained by a Z_2 projection of the N=2 manifolds or $Z_2 \times Z_2'$ projection of the N=4 manifold.

The N=2 Kahler manifold of the N=2 vector multiplet "disintigrates" as follows[35,36]

$$\frac{SO(2, \dim G_A + \dim G_B)}{SO(2) \times SO(\dim G_A + \dim G_B)} \rightarrow \frac{SO(2, n_2 n_3 + n_0 n_1)}{SO(2) \times SO(n_2 n_3 + n_0 n_1)} \tag{4.18}$$

The quaternionic manifold of the N=2 hypermultiplets decomposes according to

$$\frac{SO(4, n_A n_B)}{SO(4) \times SO(n_A n_B)} \rightarrow \frac{SO(2, n_0 n_3 + n_1 n_2)}{SO(2) \times SO(n_0 n_3 + n_1 n_2)} \times \frac{SO(2, n_0 n_2 + n_1 n_3)}{SO(2) \times SO(n_0 n_2 + n_1 n_3)} \tag{4.19}$$

The Kahler potential takes the form

$$J = \sum_{a=0}^{2} J_a(y_i) \tag{4.20}$$

where

$$J_a = -\log\left(1 - \left| y_a^\alpha \right|^2 + \frac{1}{4} \left| y_a^{\alpha 2} \right|^2\right) \tag{4.21}$$

is a "canonical" form[40] of the $\frac{SO(2, n)}{SO(2) \times SO(n)}$ Kahler potential.

The superpotential is tri-linear and coupled the three different manifolds[35,36]

$$T_{\alpha\beta\gamma}\, y_0^\alpha y_1^\beta y_2^\gamma \tag{4.22}$$

where α, β, γ are gauge indices and $T_{\alpha\beta\gamma}$ are suitable $SO(n_i)$ temor coefficients. The gauge-kinetic function is simply given by

$$f_{AB} = \delta_{AB}\, S \tag{4.23}$$

and is the same as in Calabi-Yau type compactifications.

The manifold structure and the knowledge of the basic functions of the standard N=1 supergravity theory completely defines the low-energy theory for superstring models in which the twisted sector is massive.

These models do not have ordinary fermion families. A massless twisted sector for N=2 superstring models can only exist[8] if n_A or n_B = 8 or 16, while for N=1 superstrings it can occur if $n_0 + n_i$ = 8 or 16 for at least one i.

When a massless twisted sector occurs in the theory, the low-energy supergravity Lagrangian cannot be obtained by a N=4 truncation. However, in the case of N=2 models, the only arbitrariness lies in the structure of the quaternionic manifold of the massless hypermultiplets. Once the quaternionic manifold is known, the low-energy theory is then completely fixed. For N=1 theories the situation is even more complicated because there are more functions to be determined. However the N=1 theory can be obtained by a Z_2 truncations of the N=2 theory in the case in which the N=1 massless "twisted" chiral multiplets are obtained by a Z_2 projection of the N=2 twisted hypermultiplets.

To give just an example, this situation occurs in the model obtained by the following Z_2 and $Z_2 \times Z_2'$ projections:

$$SO(44) \xrightarrow[Z_2]{} SO(16) \times SO(28) \xrightarrow[Z_2 \times Z_2']{} SO(10) \times SO(6) \times SO(14)^2. \qquad (4.24)$$

The N=1 model contains sixteen chiral families in the spinorial representation of SO(10) with gauge quantum numbers $(16, 4 + \bar{4}, 1, 1)$.

The knowledge of the twisted state interactions can be obtained by intergrating out the massive states which are coupled to the massless ones[37].

This integration can determine the structure of the quaternionic manifold of the N=2 theory and by a further Z_2 projection the structure of the Kahler manifold of the N=1 chiral theory alluded before.

We cannot give here the details. The integration over the massive string states has the effect of "deforming" the hypermultiplet quaternionic manifold is a non-symmetric, non-homogeneous space which is an SU(2)-quotient[41] of the projective quaternionic space

$$HP^n = \frac{Sp(4 + n)}{Sp(1) \times Sp(3+n)} \qquad (4.25)$$

where n is the total number of twisted and untwisted hypermultiplets. If we set the twisted hypermultiplet to vanish then this SU(2) quotient gives back the Wolf-manifold given by equation (4.9). The N=1 theory, obtained by Z_2 truncation of this SU(2) quotient, has a Kahler manifold of the form[37]

$$SU(1,1) \times \frac{SO(2, n_0)}{SO(2) \times SO(n_0)} \times \mathcal{M} \qquad (4.26)$$

when $\mathcal{M}$ is a non-symmetric, non-homogeneous Kahler manifold with Kahler potential given by[37]

$$J_{\mathcal{M}} = -2\log \left(e^{-(J_1+J_2)/2} - |z|^2\right) \qquad (4.27)$$

where J_1, J_2 are given by equation (4.21) and y_1, y_2 are the scalar fields coming from the truncation of the N=2 untwisted hypermultiplets. "z" denotes the scalar fields of the twisted chiral multiplets. The superpotential for the twisted sector is of the form

$$(ZZY_0) \qquad (4.28)$$

as it comes from the Z_2 projection of the N=2 "Yakawa couplings".

If $z = 0$, $J_{\mathcal{M}}$ reduces to a pair of $\frac{SO(2, n)}{SO(2) \times SO(n)}$ manifolds, while if we set $y_1 = y_2 = 0$, $J_{\mathcal{M}}$ reduces to a non compact CP^n manifold.

The theory discussed here is the first example of a low-energy theory of a four-dimensional superstring which includes chiral families. The N=1 scalar potential is positive semi-definite, as it can be shown from its relation with the N=2 potential[38]. However, all extrema do not break supersymmetry. An important issue is whether these models can undergo supersymmetry breaking.

Two possibilities come to mind[42]. The first is a "condensation mechanism"[43] in the $SO(14)^2$ hidden sector, in analogy with the gaugino condensation mechanism of the $E_6 \times E_8$ models[44].

A more interesting possibility would be through a Scherk-Schwarz mechanism[45] with a non-trivial compactifications of a five-dimensional superstring theory, in which the fields pick up an extra-dependence from the fifth coordinate, then breaking the supersymmetry in D=4 dimension.

A third alternative, in some cases related to the previous one, would be, to gauge the graviphotons of the N=4 theory in a way which breaks all supersymmetries in a flat background. If the gauging is consistent with the Z_2 projection, this could give rise to a N=2 or N=1 theory with broken supersymmetry.

It is clear that it may be difficult to show that any of the envisaged mechanisms do occur in the underlying string theory; however, they could be used as a "phenomenological" parametrization of supersymmetry breaking in the low-energy effective theory which in turn may be proven to be a useful tool for the application of these models to elementary particle physics.

This is an updated version of lectures given at the 1986 International School of Subnuclear Physics - Erice (Sicily)

REFERENCES

1. M.B. Green and J.H. Schwarz, Phys. Lett. 149B (1984) 117.

2. For a recent status on superstring theory see, e.g., M. Green, J.H. Schwarz, E. Witten, Superstring Theory I and II, Cambridge University Press, 1987.
M. Green, D.J. Gross, Unified String Theories, World Scientific, Singapore 1985.

3. P. Candelas, G. Horowitz, A. Strominger, and E. Witten, Nucl. Phys., B258 (1985) 46.

4. L. Dixon, J.A. Harvey, C. Vafa, E. Witten, Nucl. Phys. B261 (1985) 678; B274 (1986) 285.

5. K.S. Narain, Phys. Lett. 196B (1986) 418.

6. K.S. Narain, M.H. Sarmadi, E. Witten, Nucl. Phys. B279 (1987) 369.

7. H. Kawai, D.C. Lewellen, S.H.H. Tye, Phys. Rev. D34, (1986) 3794; Phys. Rev. Lett. 57, (1986) 1832; Cornell Preprints CLNS 87/751; 87/760 (1987).

8. I. Antoniadis, C. Bachas, C. Kounnas, Ecole Polytecnique, preprint A761.12, LBL-22709 (Dec-86).

9. W. Lerche, D. Lüst, A.N. Schellekens, CERN-TH-4590/86.

10. D.J. Gross, J.A. Harvey, E. Martinez, R. Rohn, Phys. Rev. Lett. 54 (1985) 502.

11. For a review see i.e. P.H. Nilles, Phys. Report Val 110, N.1,2 pg.1 (1984).

12. E. Witten, Phys. Lett. 155B (1985) 151.

13. M. Duff, C. Pope, B. Nilsson, N.P. Warner, Phys. Lett. 149B (1984) 90.

14. S. Ferrara, C. Kounnas, M. Porrati, Phys. Lett. B181 (1986) 263.

15. E. Cremmer, S. Ferrara, C. Kounnas, D.V. Nanopoulos, Phys. Lett. 133B (1983) 61, J. Ellis, C. Kounnas, and D.V. Nanopoulos, Nucl. Phys. B241 (1984) 406; and Nucl. Phys. B247 (1984) 373.

16. E. Cremmer, S. Ferrara, J. Scherk, Phys. Lett. 74B (1978) 61; E. Bergshoeff, I.G. Koh, E. Sezgin, Phys. Lett. 155B (1985) 71; M. DeRoo, P. Wagemans, Nucl. Phys. B262 (1985) 644.

17. F. Bergshoeff, M. deRoo, B. de Wit, P. van Nieuwenhisum, Nucl. Phys. B195, (1982) 97. G. Chapline, N. Manton, Phys. Lett. 120B (1983) 105. A.H. Chamseddine, Phys. Rev. D24 3065 (1981); Nucl. Phys. B185, (1981) 403.

18. H. Kawai, D.C. Lewellen, S.H.H Tye, CLNS 87/760, Jan 87; L. Dixon, V.S. Kaplunowsky, C. Vafa, SLAC-PUB-4282/ HUTP-87/A034 March 1987.

19. M. Dine, N. Seiberg, E. Witten, IAS Preprint (1986). M. Dine, J. Ichinose, N. Seiberg, IASSNS-HEP-87/17 Preprint March 87. J.J. Atick, L.J. Dixon, A-Sen, SLAC-PUB-4235-Feb 87.

20. E. Cremmer, S. Ferrara, L. Girardello, A. van Proeyen, Nucl. Phys. B212 (1983) 413.

21. R. Barbieri, E. Cremmer, S. Ferrara, Phys. Lett. 63B (1985) 143.

22. C.P. Burgess, A. Font, F. Quevedo, Nucl. Phys. B272 (1986) 661.

23. H.P. Nilles in "Supersymmetry, Supergravity and Superstring", Eds: B. deWit, P. Fayet, M. Grisaru (ICTP Trieste) World Scientific, Singapore (1986).

24. T. Kugo, S. Uehara, Nucl. Phys. B226 (1983) 49; Progr. Theor. Phys. 73 (1985) 253.

25. S. Cecotti, S. Ferrara, L. Girardello, M. Porrati, Phys. Lett. 185B (1987) 345.

26. For a review of supergravity see e.g. P. van Nieuwenhuizen, Phys. Rep. Vol. 68, n. 4 pg. 192 (1981)

27. S. Ferrara, B. Zumino, J. Wess, Phys. Lett. 51B (1974) 239. W. Siegel, Phys. Lett. 85B (1979) 337; V. Lindström, M. Rocek, Nucl. Phys. B222 (1983) 285; S. Ferrara, L. Girardello, T. Kugo, A. van Proeyen, Nucl. Phys. B223 (1983) 191.

28. S. Ferrara, M. Villasante, Phys. Lett. 186B (1987) 85.

29. S. Cecotti, S. Ferrara, M. Villasante, UCLA/87/TEP/12.

30. For superspace curved geometry see e.g.: J. Bagger, J. Wess, Supersymmetry and Supergravity Princeton University Press 1983-For

Chem-Simons Couplings in 4D see: S. Cecotti, S. Ferrara, L. Girardello, M. Porrati, Phys. Lett. 164B (1985) 46; G. Girardi, R. Grimm, Annecy Report, LAPP-TH-180 (1980).

31. S. Cecotti, S. Ferrara, L. Girardello, CERN-TH-4698 (March 87)-UCLA/87/TEP/5.

32. S. Cecotti, S. Ferrara, L. Girardello, Phys. Lett. 187B (1987) 321.

33. M. Muller, E. Witten Phys. Lett. 182B (1986) 28.

34. I. Antoniadis, C. Bachas, C. Kounnas, P. Windey, Phys. Lett. 171B (1986) 51.

35. S. Ferrara, L. Girardello, C. Kounnas, M. Porrati, LBL-22905/UCLA/87/TEP/2-ULB/TPH/87/4 to appear in Phys. Lett. B.

36. I. Antoniadis, J. Ellis, E. Floratos, D.V. Nanopoulos, T. Tomaras, CERN-TH 4651/87-MAD/TH/87-02

37. S. Ferrara, L. Girardello, C. Kounnas, M. Porrati, UCLA/87/TEP/11. LBL 23335, to appear in Phys. Lett. B.

38. B. de Wit. R.G. Lawers, A. van Proeyen, Nucl. Phys. B255 (1985) 569; E. Cremmer et al. Nucl. Phys. B250 (1985) 385; E. Cremmer, A. van Proeyen, Class Quantum Grav. 2 (1985) 445.

39. J. Bagger, E. Witten, Nucl. Phys. B222 (1983) 1; J.A. Wolf, J. Math. Phys. 14 (1965) 1033.

40. E. Calabi, E. Vesentini, Annals of Math Vol. 71, no.3 (1960) 472.

41. K. Galicki, Comm. Math. Phys. 108 (1987), K. Galicki, Stony Brook preprint, ITP-SB 86,96 (1986).

42. S. Ferrara, C. Kounnas, M. Porrati, F. Zwirner, LBL-23361 preprint UCLA/87/TEP/15-UCB/TPH/87/19, to appear in Phys. Lett. B.

43. S. Ferrara, L. Girardello, P.H. Nilles, Phys. Lett. 125B, (1983) 457

44. M. Dine, R. Rohm, N. Sieberg, E. Witten, Phys. Lett. 156B (1985) 55.

45. J. Scherk, J.H. Schwarz, Phys. Lett. 82B (1979) 60; Nucl. Phys. B152 (1979) 61; E. Cremmer, J. Scherk, J.H. Schwarz, Phys. Lett. 84B (1979) 83; J.H. Schwarz, Phys. Lett. 95B (1980) 219

DISCUSSION

CHAIRMAN: S. Ferrara

Scientific Secretaries: G. Miele, A. Pasquinucci, S. Haywood and J. Quackenbush

– *Y. Shamir:*

You have discussed super gravity models for which the scalar potential is positive semi–definite after spontaneous SUSY breaking, and you mentioned that there are models for which the minimum of the potential is zero. Is there a symmetry principle that is involved in it? Do all the arguments apply to non–perturbative effects as well?

– *S. Ferrara :*

Under general assumptions in N=1 SUGRA theories, you can imagine a condition for which the potential is semi–positive definite for arbitrary configurations of the scalar field. The chiral potential can be written as:

$$e^{G}[G^{i}(G^{-1})_{i}^{j}\, G_{j} - 3]$$

If this function is semi–positive definite, there are two possible situations. In one case, there is unbroken SUSY with zero vacuum energy.It is a situation in which

$$< e^{\frac{G}{2}} G_i > = 0$$

This term is connected to the scalar shift in the fermions. The gravitino mass also vanishes, which means

$$e^{\frac{G}{2}} = 0$$

There is another possibility where we have

$$< e^{\frac{G}{2}} G_i > \neq 0.$$

Since the potential is positive definite, this means that the gravitino mass is also non–zero, so

$$e^{\frac{G}{2}} \neq 0.$$

This is the reason why you can have broken SUSY with zero vacuum energy in SUGRA. This is impossible for the Lagrangian of 10–D SUGRA, because of the presence of two scalar fields s and t, which are the two Wess–Zumino multiplets. The real part of these fields are related to the dilation ϕ and the compactification modes which come from 10–D string theories.

In a more complicated model, you can have more scalar fields, for instance in orbifolds, but you have at least two scalar fields.

If you relax the 10–D SUGRA Lagrangian to allow a potential which depends on the s field in a non–trivial (as in the case of possible gluino condensation analysed by Dine, Witten, Seiberg and Rhom), then the s field can be responsible for the breaking, and in this case the gravitino can have an arbitrary non–zero mass at the tree level (the t multiplet is actually the Goldstino multiplet).

It is not clear if this kind of semi–positive potential is contained in superstring theory. If you choose the Kähler potential so that

$$D(\phi_i, \overline{\phi}_j) = det \left(- \frac{\partial^2 \Phi}{\partial \phi_i \partial \overline{\phi}_j} \right) \leq 0$$

you can always get a semi–positive potential. All this is based on my purely classical Lagrangian. However, the question of whether the minimum is zero is model dependent. No symmetry argument is known, and there is no reason why non–perturbative effects cannot induce $V(\Phi) > 0$.

The structure of the scalar potential is

$$V(\phi) = e^G[G^i(G^{-1})_i^j G_j - 3]$$

There is sufficient condition for the scalar potential to be positive semi–definite which was discussed in the lecture, namely:

$$D(\Phi, \overline{\Phi}) \equiv det \left. \frac{-\partial^2 \Phi}{\partial \phi_i \partial \overline{\phi}_j} \right|_{\Phi, \overline{\Phi}} \leq 0$$

$$\Phi = e^{-G/3}$$

Also

$$V(\Phi\overline{\Phi}) = 0 \Longleftrightarrow D(\Phi\overline{\Phi}) = 0$$

For string theory generated Φ one has the general structure

$$\Phi = (t + t^* - cc^*)(s + s^*)^{\frac{1}{3}} |W(c)|^{-2/3}$$

where

$$t + t^* - cc^* = e^{-\phi} e^{\sigma}$$

$$s + s^* = e^{\phi} e^{3\sigma}$$

for trivial W (say $W = 1$) $D(\Phi\overline{\Phi})$ is identically zero and therefore the scalar potential is also identically zero.

Witten et al. have shown that if W depends on s then the minimum of $V(\phi_i)$ cannot occur at $W(\phi) = 0$, namely the minimum will break SUSY.

– *F. Quevedo:*

The Lagrangian of N=1 SUGRA coupled to matter depends on two arbitrary functions when terms in the Lagrangian, up to second derivatives, are considered. In the case of higher derivative SUGRA, what is the situation?

– *S. Ferrara:*

In principle, for any given order in the derivatives, you can write all the possible functions which correspond to every invariant SUSY term, which you can construct.

In the case of a Lagrangian which contains only two derivatives, these are simply the kinetic terms of the Yang–Mills fields, and what I called the G–function is a combination of the Kähler and super–potentials.

If you include R^2 terms, then another analytic function will appear in the Lagrangian. This function is a scalar analytic function. In the case of string theories, this function is specified, at least if you take the usual results, to be linear in the s field. If you include the higher order terms, the analysis is more complicated, but in principle you have the ingredients to classify the invariants one can construct. At zero string loop you expect that all higher order terms (with higher derivatives) will also be given in terms of the basic G function that is present in the two–dirivative term. This is a consequence of the Kähler geometry and it was proven, under some general assumptions, by Cecotti, Girardello, Porrati and myself (Phys. Lett. **B185** (1987) 345). However, higher string loops generally modify this result having some unespected stringy non–renormalization theorems.

– *M. McGuigan:*

Can the twisted sectors of string theory explain why $SU(3) \times SU(2) \times U(1)$ is an anomaly free representation?

– *S. Ferrara:*

In the example I gave with the $T6/Z3$ orbifold, the twisted sector is simply what you get from trivial dimensional reduction of 10–D SUGRA when you keep not only the $SU(3)$ singlet, but also the $Z3$ singlet. The charged fields are 3 copies of the (3,27) representation of $SU(3) \times E_6$. The untwisted sector is quantum mechanically inconstistant because of $SU(3)$ gauge anomalies.

There are two classes of Wess–Zumino multiplets in the twisted sector. There are 81 in $(\overline{3}, 1)$ and 27 in (1,27). There are 27new E_6 families which are anomaly free. The 81 singlets, which is trivial under $SU(3)$ are just the objects which cancel the anomaly of the charged families.

Any orbifold has this property that there is a cancellation of the anomalies between the twisted and untwisted sectors which come automatically from string theories because of modular invariance.

– *M. Quiros:*

Do you think that the no–scale structure could be preserved by the compactification of the Chern–Simons terms + supersimmetric extensions?

– *S. Ferrara:*

If you include the R^2 terms which are necessary if you require consistency with the 10–D Bianchi identity

$$dH = F\Lambda F - R\Lambda R$$

in terms of 4–D fields, the R^2 term induces modification of the scalar potential which, however doesn't affect the extreme of the theory. So the R^2 term does not introduce "run away" solutions, if you look at the theory from a 4–D point of view.

There is another interesting fact. The R^2 (as well as the F^2 term) breaks the Kähler invariance of the Kähler potential, but only in the s–direction as the s–field is the only field coupled to the R^2 term. The resulting theory then only has Kähler symmetry in the direction of the other fields.

– *M. Carter:*

What is the content of the Gibbons–Friedman–West theorem?

– *S. Ferrara:*

Previously, in string theory, there was a theorem by Gibbons–Friedman–West which excluded compactification of 10–D Kähler SUGRA in $M_4 \times K_6$. These solutions were excluded because compactification is induced by non–zero vacuum expectation of the F–curvature. If $< F > \neq 0$, the dilaton equation of motion is not consistent with a 4–D Minkowski background. If you add the R^2 term to the Lagrangian, the expectation value of the R–curvature can match that of the F–curvature so that the equation of motion of the dilaton is compatible with the 4–D Minkowski background. You can entirely justify this mechanism from the point of view of 10–D anomaly cancellation in SUGRA.

– *K. Meissner:*

What are the arguments that supergravity is perturbatively non–renormalizable?

– *S. Ferrara:*

It is the same argument that tells you that Einstein gravity is non–renormalizable. In pure SUGRA theory, one can show that one loop and two–loop finitenesses are possible.

One–loop finiteness is achieved by the generalization of the Gauss–Bonnet theorem to SUGRA. On the two–loop level, there a miracle occurs. There are no SUSY R^3 terms, because of on–shell SUSY, so SUGRA is finite. At the 3–loop level, there are R^4 terms whose coefficients are expected to be non–zero because they are not protected by any symmetry. This is the reason why SUGRA may not be finite.

– *P. Vecsernyes:*

Could you comment in more detail on what difficulties arise when the scalar potential has negative values for some fields configurations?

– *S. Ferrara:*

Again, let me comment on the question of having vanishing vacuum energy, at least in the classical limit. In supestring theories this is not a property of superstring theories, but it is a property of 10–D SUGRA. The reason why we have this no–scale structure and vanishing vacuum energy in 10–D SUGRA is related to the existance of the dilaton field. This is the crucial difference between 10–D and 11–D SUGRA. The compactification of 11–D SUGRA is completely different from that of 10–D SUGRA. The fact that you get a configuration with zero vacuum energy in 10–D is due to the equation of motion of the dilaton, which is absent in 11–D. In these theories with semi–positive definite potentials, the only SUSY vacua are Minkowski vacua. If you have a theory where the potential can also be negative, you can have anti–De Sitter vacua.

In these theories where ou can have both Minkowski and anti–De Sitter vacua, which are both completely stable because of SUSY, you have two vacua which are good from the point of view of the theory. However, it is much better to have a solution in which you can simulate reality, namely to have a Minkowski vacuum instead of anti–De Sitter background. I don't know whether you can have an anti–De Sitter background in string theories when you include all the loop effects.

HETEROTIC SUPERSTRINGS

David J. Gross

Department of Physics
P.O. Box 708
Princeton University
Princeton, New Jersey 08544, U.S.A.

INTRODUCTION

High energy physics is, at present, in an unusual state. It has been clear for some time that we have succeeded in achieving many of the original goals of particle physics. We have constructed theories of the strong, weak and electromagnetic interactions, have understood the basic constituents of matter and their interactions. The standard model has been remarkably successful and seems to provide an accurate and complete description of physics, at least at energies below a Tev. Indeed, as we have heard in the experimental talks at this meeting, there are at the moment no significant experiments that cannot be explained by the color gauge theory of the strong interactions (QCD) and the electro-weak gauge theory. New experiments continue to confirm the predictions of these theories and no new phenomenon have appeared.

This success has not left us sanguine. Our present theories contain too many arbitrary parameters and unexplained patterns to be complete. They do not explain the dynamics of chiral symmetry breaking or of CP symmetry breaking. The strong and electroweak interactions cry out for unification. Finally we must ultimately face up to the task of including quantum gravity within the theory. However, we theorists are in the unfortunate situation of having to address these questions without the aid of experimental clues. Furthermore, extrapolation of present theory and early attempts at unification suggest that the natural scale of unification is 10^{16} Gev or greater, tantalizingly close to the Planck mass scale of 10^{19} Gev. It seems very likely that the next major advance in unification will include gravity. I don't mean to suggest that new physics will not appear in the range of Tev energies. Almost all attempts at unification do, in fact, predict a multitude of new particles and effects that could show up in the Tev domain - Higgs particles, Supersymmetric partners, etc - whose discovery and exploration is of the utmost importance. But the next truly new threshold might lie in the totally inaccessible Planckian domain.

In these lectures I shall review the present status of an ambitious attempt to construct unified theories of everything based on strings. As you are well aware,this field has seen remarkable developments in the last two years. I shall briefly review the structure of the heterotic string theory and the attempts to reconstruct the standard model from it. Then I will give a progress report on recent developments in string theory.

STRING THEORIES

String theories offer a way of realizing the potential of supersymmetry, Kaluza-Klein unification and much more. They represent a radical departure from ordinary quantum field theory, but in the direction of increased symmetry and structure. They are based on an enormous increase in the number of degrees of freedom of the world. In addition to fermionic coordinates and extra dimensions, the basic entities are extended one dimensional objects instead of points. Correspondingly the symmetry group of nature is greatly enlarged, in a way that we are only beginning to comprehend. At the very least this extended symmetry contains the largest group of symmetries that can be contemplated within the framework of point field theories – those of 10 dimensional supergravity and super Yang-Mills theory.

The origin of these symmetries can be traced back to the geometrical invariance of the dynamics of propagating strings. Traditionally, string theories are constructed by the first quantization of a classical relativistic one dimensional object, whose motion is determined by requiring that the invariant area of the world sheet it sweeps out in space-time is extremized.

In this picture the dynamical degrees of freedom of the string are the coordinates, $X^{\mu}(\sigma,\tau)$ (plus fermionic coordinates in the case of the superstring), which describe its position in space time. The symmetries of the resulting theory are all consequences of the reparametrization invariance of the σ,τ parameters which label the world sheet. As a consequence of these symmetries one finds that the free string contains massless gauge bosons . The closed string automatically contains a massless spin two meson, which can be identified as the graviton, whereas the open string, which has ends to which charges can be attached, yields massless vector mesons, which can be identified as Yang-Mills gauge bosons.

String theories are inherently theories of gravity. Unlike ordinary quantum field theory, however, we do not have the option of turning off gravity. The gravitational, or closed string, sector of the theory must always be present for consistency, even if one starts by considering only open strings, since these can join at their ends to form closed strings. One could even imagine discovering the graviton in the attempt to construct string theories of matter. In fact, this was the historical course of events, whereby the graviton (then called the Pomeron) was discovered in the dual resonance model as a bound state of open strings. Most exciting is that string theories provide for the first time a consistent, even finite, theory of gravity. The problem of ultraviolet divergences is bypassed in string theories which contain no short distance infinities. This is not too surprising considering the extended nature of strings, which softens their interactions. Alternatively one notes that interactions are introduced into string theory by allowing the string coordinates, which are two dimensional fields, to propagate on world sheets with nontrivial topology that describe strings spliting and joining. ¿From this point of view, based on first quantization, one does not introduce an interaction at all, one just adds handles or holes to the world sheet of the free string. As long as reparametrization invariance is maintained there are simply no possible counterterms. In fact all the divergences that have ever appeared in string theories can be traced to infrared divergences that are a consequence of vacuum instability. This comes about because all string theories contain a massless partner of the graviton called the dilaton. If one constructs a string theory about a trial vacuum state in which the dilaton has a nonvanishing vacuum expectation value, then infrared infinities will occur due to massless dilaton tadpoles. These divergences however are just a sign of the instability of the original trial vauuum. This is the source of the divergences that occur in one loop diagrams in the old bosonic string theories (the Veneziano model). Superstring theories have vanishing dilaton tadpoles, at least to one-loop order. Therefore both the superstring and the heterotic string are explicitly finite to one loop order and there are strong arguments that this persists to all orders!

String theories, as befits unified theories of physics, are incredibly unique. In principle they contain no freely adjustable parameters and all physical quantities should be calculable in terms of h, c, and m_{planck}. In practice we are not yet in the positon to exploit this enormous predictive power. The fine structure constant α, for example, appears in the theory in the form $\alpha exp(-D)$, where D is the aforementioned dilaton field. Now the value of this field is undetermined to all orders in perturbation theory (it has a flat potential). Thus we are free to choose its value, thereby choosing one of an infinite number of degenerate vauum states, and thus to adjust α as desired. Ultimately we might belive that string dyamics will determine the value of D uniquely, presumably by a nonperturbative mechanism, and thereby eliminate the nonuniqueness of the choice of vacuum state. In that case all dimensionless parameters will be calculable. Even more string theories determine in a rather unique fashion the gauge group of the world and fix the number of space-time dimensions to be ten.

Finally and most importantly, string theories lead to phenomenologically attractive unified theories, which could very well describe the real world.

CONSISTENT STRING THEORIES

The number of consistent string theories is extremely small, the number of phenomenologically attractive theories even smaller. First, there are the closed superstrings, of which there are two consistent versions. These are theories which contain only closed strings which have no ends to which to attach charges and are thus inherently neutral objects. At low energies, compared to the mass scale of the theory which we can identify as the Planck mass, we only see the massless states of the theory which are those of ten dimennsional supergravity. One version of this theory is non-chiral and of no interest since it could never reproduce the observed chiral nature of low energy physics. The other version is chiral. One might then worry that it might suffer from anomalies, which is indeed the fate of almost all chiral supergravity theories in ten dimensions. Remarkably the particular supergravity theory contained within the chiral superstring is the unique anomaly free theory in ten dimensions. It however contains no gauge interactions in ten dimensions and could only produce such as a consequence of compactification. This approach raises the same problems of reproducing chiral fermions that plagued field theoretic Kaluza-Klein models and has not attracted much attention.

Open string theories, on the other hand, allow the introduction of gauge groups by the time honored method of attaching charges to the ends of the strings. String theories of this type can be constructed which yield, at low energies, $N = 1$ supergravity with any Yang-Mills group. These however , in addition to being somewhat arbitrary, were suspected to be anomalous. The discovery by Green and Schwarz that, for a particular gauge group SO_{32}, the would be anomalies cancel, greatly increased the phenomenological prospects of unified string theories.

The anomaly cancellation mechanism of Green and Schwarz also provided the motivation that led to the discovery of a new string theory, whose low energy limit contained an E_8XE_8 gauge group–the heterotic string. The heterotic string is a closed string theory that produces, by a stringy generalization of Kaluza-Klein compactification, gauge interactions. These are required by consistency to be either E_8XE_8 or $Spin_{32}/Z_2$. The E_8XE_8 version of this theory offers the best phenomenological prospects for reproducing the real world. In fact the group E_8 was explored seriously at as a GUT group by theorists who extrapolated upward from the standard model, so one might hope to be able to proceed in the opposite direction.

THE HETEROTIC STRING

Previously known string theories are the bosonic theory in 26 dimensions (the Veneziano model) and the fermionic, superstring theory in 10 dimensions (an outgrowth of the Ramond-Neveu-Schwarz string). The new string theory is constructed as a chiral hybrid of these. To see how this is possible let us recall how string theories are constructed.

Free string theories are constructed first quantization of an action given by the invariant area of the world sheet swept out by the string, or by its supersymmetric generalization. For the bosonic string the action is

$$S = -T \int d\sigma d\tau [\eta_{\alpha\beta} g^{ab} \partial_a X^\alpha \partial_b X^\beta], \tag{1}$$

where $X^\mu(\sigma,\tau)$ labels the space time position of the string, embedded in some D dimensional manifold ($\mu = 1, 2, ..., D$), with σ, τ labeling the world sheet that the string sweeps out. T is the string tension which, for the heterotic string, is essentially equal to m^2_{Planck}. It appears to be possible to construct consistent string theories as long as the above two dimensional σ -model is conformally invariant. For the moment we take the big space to be flat, so that $\eta_{\alpha\beta}$ is the Minkowski metric. This is essentially a choice of vacuum for the quantum string theory. In order to describe the real world however one will be interested in non flat D dimensional manifolds. The reparametrization invariance of the action (in σ, τ) permits one to choose the metric, g_{ab} of the world sheet to be conformally flat and in which the timelike parameter of the world sheet, τ, is identified with light cone time. In this light cone gauge the theory reduces to a two-dimensional free field theory of the physical degrees of freedom – the transverse coordinates of the string, subject to constraints. This procedure is valid however only in the critical dimension of 26 for the bosonic string and 10 for the fermionic string. In other dimensions of space time the existence of conformal anomalies imply that the conformal degree of freedom of the internal metric does not decouple. If it is ignored there is a breakdown of world sheet reparametrization invariance.

In the critical dimension the physical degrees of freedom, being massless two-dimensional fields, can be decomposed into right and left movers, i.e. functions of $\tau - \sigma$ and $\tau + \sigma$ respectively. If we consider only closed strings then the right and left movers never mix. This separation is maintained even in the presence of string interactions, as long as we allow only orientable world sheets on which a handedness can be defined. This is because the interactions between closed string are constructed, order by order in perturbation theory, by simply modifying the toplogy of the world sheet on which the strings propagate. In terms of the first quantized two-dimensional theory no interaction is thereby introduced; the right and left movers still propagate freely and independently as massless fields. Thus there is, in principle, no obstacle to constructing the right and left moving sectors of a closed string in a different fashion, as long as each sector is separately consistent and together can be regarded as a string embedded in ordinary space-time. This is the idea behind the construction of the heterotic string, which combines the right movers of the fermionic superstring with the left movers of the bosonic string. It is necessarily a theory of closed and orientable strings, since one can clearly distinguish an orientation on such a string. In some sense the heterotic string is inherently chiral; indeed we do not have the option, present in other closed string theories, of constructing a left-right symmetric theory.

The physical degrees of freedom of the right-moving sector of the fermionic superstring consist of 8 transverse coordinates $X_i(\tau - \sigma)$ (i=1,...8) and 8 Majorana-Weyl fermionic coordinates $S_a(\tau - \sigma)$. The physical degrees of freedom of the left-moving sector of the bosonic string consist of 24 tranverse coordinates, $X_i(\tau + \sigma)$ and $X_I(\tau - \sigma)$ (i=1,...8, I=1,....16). Together they comprise the physical degrees of freedom of the heterotic string. The eight transverse right and leftmovers combine with the longitudinal coordinates to describe the position of the string embedded in 10 dimensional space. The extra fermionic and bosonic degrees of freedom parametrize an internal space.

The light cone action that yields the dynamics of these degrees of freedom can be derived from a manifestly covariant action, and one can easily quantize it. The only new feature that enters is the compactification and quantization of the extra 16 left-moving bosonic coordinates. It is this compactification, on a uniquely determined 16 dimensional compact space, that leads to the emergence of Yang-Mills interactions.

The extra 16 left moving coordinates of the heterotic string can be viewed as parametrizing an "internal" compact space T. This interpretation should not be taken too literally, in fact one can equally well represent these degrees of freedom by 32 real fermions. The question then arises as to the nature of the internal manifold T. This should be a dynamical question since the string theory is a theory of gravity, thus the choice of a backround spacetime is a dynamical issue–the background must be a solution of the string equations of motion. The remarkable feature of the heterotic string is that the space T is completely determined to be a very special torus (the maximal torus of E_8XE_8 or $Spin_{32}/Z_2$.

That a torus is a solution is reasonable, since a torus is simply a flat space with periodic boundary conditions; however, there are many 16 dimensional tori. What picks out the special torus is the fact that the coordinates of T are leftmoving only (i.e.. functions of $\tau+\sigma$. Consider then the expansion of the internal coordinates $X_I(\tau+\sigma)$

$$X_I(\tau+\sigma)=X_I+P_I\tau+L_I\sigma+\text{oscillators}. \tag{2}$$

The momentum P_I is quantized (since X_I lives on a compact domain) in units of $\frac{1}{R}$ (where R is a radius of T). The term L_I must go around the torus (in some direction) an integer number of times so that X_I be a periodic function of σ. Therefore L_I must equal an integer multiple of the radius of T in some direction. A string configuration with nonvanishing L_I represents a soliton, i.e. a string that winds around the torus some number of times, and the winding number is a toplogical, conserved charge. Since X_I is a function of $\tau+\sigma$, L_I must be to equal P_I . This clearly restricts the form of the torus. It obviously means that $R=1$ (in our units this means $\frac{1}{M_{Planck}}$.)

The form of T is further constrained since for a general torus it is not possible to identify the winding numbers,(which span a lattice Γ , $T=R_{16}/\Gamma$), with the momenta, P_I , which lie on the lattice Γ^* dual to Γ . In fact, the full consistency of the heterotic string theory requires that T be such that $\Gamma=\Gamma^*$, i.e. that the lattice defining the torus be self dual. If this is not satisfied then "modular invariance" breaks down. This means that the amplitudes, at the one loop level, develop anomalies which destroy reparametrization invariance. Now there exist very few self dual lattices of the appropriate type, in fact they only exist in 8N dimensions! In 8 dimensions there is Γ_8 , which coincides with the root lattice of E_8XE_8 ; and in 16 dimensions there are two self dual lattices, $\Gamma_8X\Gamma$ and Γ_{16} , where Γ_{16} is the weight lattice of $Spin_{32}/Z_2$. Thus modular invariance, i.e. invariance of the two dimensional world sheet under global diffeomorphisms, is responsible for the emergence of these two gauge groups.

Let us now examine the massless particles of the heterotic string. It contains, of course, gravity, and since it possesses one supersymmentry (that transforms the supersymmetric right movers) it will contain the massless multiplet of ten dimensional, N=1, supergravity. We would also expect that, in analogy to the standard Kaluza-Klein mechanism, a compactified string theory will contain massless vector mesons associated with the isometries of the compact space. For T this would yield the 16 gauge bosons of U_1^{16}. A remarkable feature of closed string theories is that for special choices of the compact space there will exist extra massless gauge bosons, which are in fact massless solitons. These are string configurations that wind around the internal 16 torus and have non vanishing momenta=winding number $=U_1^{16}$ charge. They combine with the Kaluza-Klein gauge bosons to fill out the adjoint representation of a simple Lie group whose rank equals the dimension of T. For the two allowed choices of T these produce the gauge bosons of G= $Spin_{32}/Z_2$ or E_8XE_8.

The heterotic string theory has, by now, been developed to the same stage as other superstring theories. Interactions have been introduced, and shown to preserve the symmetries and consistency of the theory, radiative corrections calculated and shown to be finite. In many ways it now appears as the simplest of all superstring theories. It surely provides a most satisfactory explanation for the emergence of specific gauge interactions and, as we shall see, offers much phenomenological promise.

STRING PHENOMENOLOGY

In order to make contact between the string theories and the real world one is faced with a formidable task. These theories are formulated in ten flat space-time dimensions, have no candidates for fermionic matter multiplets, are supersymmetric and contain an unbroken large gauge group-say E_8XE_8. These are not characteristic features of the physics that we observe at energies below a Tev. If the theory is to describe the real world one must understand how six of the spatial dimensions compactify to a small manifold leaving four flat dimensions, how the gauge group is broken down to $SU_3XSU_2XU_1$, how supersymmetry is broken, how families of light quarks and leptons emerge, etc.. Much of the recent excitement concerning string theories has been generated by the discovery of a host of mechanisms, due to the work of Witten and of Candelas, Horowitz and Strominger, and of Dine, Kaplonovsky, Nappi, Seiberg, Rohm, Breit, Ovrut, Segre and others, which indicate how all of this could occur. The resulting phenomenolgy, in the case of the E_8XE_8 heterotic string theory is quite promising.

The first issue to be addressed is that of the compactification of six of the dimensions of space. The heterotic string, as described above, was formulated in ten dimensional flat spacetime. This however is not neccesary. Since the theory contains gravity the issue of which spacetime the string can be embedded in is one of string dynamics. That the theory can consistently be constructed in perturbation theory about flat space is equivalent to the statement that ten dimensional Minkowski spacetime is a solution of the classical string equations of motion. Such a solution yields the background expectation values of the quantum degrees of freedom. We can then ask are there other solutions of the string equations of motion that describe the string embedded in, say, four dimensional Minkowski spacetime times a small compact six dimensional manifold ?

At the moment we do not possess the full string functional equations of motion, however one can attack this problem in an indirect fashion. One method is to deduce from the scattering amplitudes that describe the string fluctuations in ten dimensional Minkowski space an effective Lagrangian for local fields that describe the string modes. Restricting ones attention to the massless modes, the resulting Lagrangian yields equations which reduce to Einstein's equations at low energies, and can be explored for compactified solutions. Another method is to proceed directly to construct the first quantized string about a trial vacuum in which the metric (as well as other string modes) have assumed backround values. In this approach one starts with the action of Equation (1), or its supersymmetric generalization, but allows $\eta_{ab}(X)$ to be the metric of a curved manifold. A consistent string theory can be developed as long as the two dimensional field theory of the coordinates $X^\mu(\sigma,\tau)$ is conformally invariant. This is a nontrivial requirement, since the theory described by (1) is an interacting nonlinear

σ -model. The condition that the two dimensional theory be conformal invariant is equivalent to demanding that the string equations of motion are satisfied. Thus one can search for alternative vacuum states by looking for σ -models (actually supersymmetric σ -models), for which the relevent β-functions (which are local functions of the metric η_{ab} and its derivitives) vanish. In addition one must check that the anomaly in the commutators of the stress energy tensor is not modified. Given such a theory one can construct a consisitent string theory and if η_{ab} describes a curved manifold the string will effectively be embedded in this manifold.

Remarkably there do exist a very large class of conformally invariant supersymmetric σ -models, that yield solutions of the string classical equations of motion to all orders and describe the compactification of ten dimensions to a product space of four dimensional Minkowski space times a compact internal six dimensional manifold. These compact manifolds are rather exotic mathematical constructs (they are Kahler and admit a Ricci flat metric– i.e. they have SU_3 holonomy). and are called 'Calabi-Yau' manifolds. In general they have many free parameters (moduli) which, among the rest, determine their size. Once again, this is an indication of the enormous vacuum degeneracy of the string theory, at least when treated perturbatively, and leads to many (at the present stage of our understanding) free parameters. This abundance of riches should not displease us, at the moment we would like to know whether there are any solutions of the theory which resemble the real world, later we can try to understand why the dynamics picks out a particular solution.

In the case of the heterotic string it is not sufficient to simply embed the string in a Calabi-Yau manifold. One must also turn on an SU_3 subgroup of the E_8XE_8 gauge group of the string. This is because the internal degrees of freedom of the heterotic string consist of right-moving fermions, which feel the curvature of space-time, and left moving coordinates which know nothing of the space-time curvaure but are sensitive to background gauge fields. Unless there is a relation between the curvature of space and the curvature (field strength) of the gauge group there is a right left mismatch which gives rise to anomalies. Therefore one must identify the space-time curvature with the gauge curvature (embed the spin connection in the gauge group). One does this by turning background gauge fields in an SU_3 subgroup of one of the E_8's, thereby breaking it down to E_6 (or possibly O_{10} or SU_5).

These Calabi-Yau compactifications, produce for each manifold K, a consistent string vacuum, for which the gauge group is no larger than E_6XE_8 and $N=1$ supersymmetry is preserved. Furthermore there now exist massless fermions which naturally form families of quarks and leptons. Recall that, after Kaluza-Klein compactification, the spectrum of massless chiral fermions is determined by the zero modes of the Dirac operator on the internal space. Since, for the heterotic string, the gauge and spin connections are forced to be equal one can count the number of chiral fermions by geoemtrical argumments. The massless fermions fall into the **27** of E_6. This is good, E_6 is an attractive grand unified model and each **27** can incorporate one generation of quarks and leptons. The number of generations is equal to half the Euler character of the manifold (which counts the number of handles it has), and is normally quite large. If there exists a discrete symmetry group, Z, which acts freely on K, one can consider the smaller manifold K/Z, whose Euler character is reduced by the dimension of Z. By this trick, and after some searching, manifolds have been constructed with 1,2,3,4,....generations. It seems that to be realistic we must restrict attention to manifolds with three, or perhaps four, generations. This compactification scheme also produces a natural mechanism for the breaking of E_6 down to the observed low energy gauge group. If K/Z is multiply connected one can allow flux of the unbroken E_6 (or of the E_8 for that matter) to run through it, with no change in the vacuum energy. The net effect is that when we go around a hole in the manifold through which some flux runs we must perform a nontrivial gauge transformation on the charged degrees of freedom. These noncontractible Wilson loops act like Higgs bosons,breaking E_6 down to the largest subgroup that commutes with all of them. By this mechanism one can, without generating a cosmological constant, find vacua whose unbroken low energy gauge group is, say, $SU_3XSU_2XU_1X$ (typically, an extra U_1 or two). Moreover there exists a natural reason for the existence of massless Higgs bosons which are weak isospin doublets (and could be responsible for the electro-weak breaking at a Tev), without accompanying color triplets. Many of the sucessful features of grand unified models, such as the prediction of the weak mixing angle, carry over, and many of the unsuccessful predictions, such as quark lepton mass ratios, do not.

Of course it is also necessary to break the remaining N=1 supersymmetry. For this purpose the extra E_8 gauge group might be useful. Below the compactification scale it yields a strong, confining gauge theory like QCD, but without light matter fields. In general this sector would be toally unobservable to us, consisting of very heavy glueballs, which would only interact with our sector with gravitational strength at low energies. However there could very well exist in this sector a gluino condensate which can serve as source for supersymmetry breaking.

Thus the heterotic string theory appears to contain, in a rather natural context, many of the ingredients necessary to produce the observed low energy physics. I do not mean to suggest that there are not many problems and unexplained mysteries. There exists the danger (common to many grand unified models, especially supersymmetric ones) of too rapid proton decay, there is no deep understanding of why the cosmological constant, so far zero, remains zero to all orders, and when supersymmetry is broken, at least by the mechanism discussed above, the theory tends to relax back to ten dimensional flat space. Nonetheless the early successes are very reassuring and they give one the feeling that there are no insuperable obstacles to deriving all of low energy physics from the E_8XE_8 heterotic string theory.

RECENT PROGRESS

I do not want to leave the impression that string theory has brought us close to the end of particle physics. Quite the opposite is the case. Not only are there many unsolved problems and deep mysteries that need to be understood before one can claim sucess, in addition we have only begun to probe the structure of these new theories. I shall present a list of problems that are the focus of current research and comment on recent progress.

What Is String Theory?

We do not fully understand the deep principles and symmetries that underly string theories. To date these theories have been constructed in a somewhat adhoc fashion and often the formulism has produced, for reasons that are not totally understood, structures that appear miraculous. There is a strong feeling among string theorists that the analogue of the principle of equivalence for this extension of general relativity has not yet been discovered. To date all treatments of string theory have been carried out using the methods of first quantization. This procedure leads to the description of string scattering amplitudes as sums of path integrals of two-dimensional σ models on world sheets with any number of handles. Unlike particle theories this formulation is very natural and beautiful (for example interactions are introduced in a geometric and unique fashion) but one might ask whether first quantization is enough? In my opinion the answer is– probably not! First, although the path integral formulation is very pretty, it does not manifestly exhibit all the symmetries of the theory; second, it only yields a perturbative expansion of the theory which surely is, at best, an asymptotic expansion of the theory ; and finally, it is a background dependent. In fact, were the first quantized perturbation theory to define the full string theory by itself we would likely be in deep trouble. This is because there are many properties of all string theories that are true to all orders in perturbation theory but not in the real world (exact supersymmetry, massless scalars, etc..).

Most likely a 'second quantized' treatment of strings is required, as in the case of ordinary field theories, where we introduce operator valued string fields, $\Phi[x^{\mu}(\sigma,\tau)]$, which are functionals of loops and a Hamiltonian that generates the dynamics. In fact, everything we know to date of the structure of string theories suggests that it is nothing more than an "ordinary" field theory, albeit one with an very large number of degrees of freedom. This is the conservative approach. It is also possible that a novel approach (such as that advocated by Friedan and Shenker, based on infinite genus Riemann surfaces) will prove more fruitful.

String field theory has been developed in light cone gauge, years ago for the bosonic (by Kaku and Kikkawa , Cremmer and Gervais, and Mandelstam)and more recently for superstrings (by Brink, Green and Schwarz) and for heterotic strings (by V.Periwal and myself). The advantage of this approach is that it is manifestly unitary, contains only physical degrees of freedom and that it exists. Its disadvantage is that, having fixed a gauge, manifest symmetry (even global Lorentz invariance) is lost and the formulation is background dependent.

Much effort has been expended recently in attempts to formulate covariant string field theory. One of the most important developments (due mainly to W.Siegel) is the realization that (first quantized) ghost coordinates are not unphysical, but rather natural devices for discussing forms on loop space. With this insight free string field theories have been formulated for both the bosonic and superstring theories. As for interactions two separate approaches have been pursued. First there is the attempt (by Siegel and Zwiebach, by Hato, Itoh, Kugo, Kunimoto and Ogawa and by Neveu and West) to graft the vertex of the light cone field theory onto a covariant approach. This direction suffers from two problems. First, it requires the introduction of an unphysical parameter, α (which in light cone gauge is identified with P_{+}), whose role is the measure the length of the string and second, it is not based on an apparant geometrical principle. A different approach has been pursued by Witten, motivated by a generalization of ordinary geometry to non-commutative geometry .

Witten constructs an abstract geometrical framework which is so that the first-quantized BRST operator, Q, plays the role of an exterior derivative (this is motivated by the fact that $Q^2 = 0$). To this end he defines a (non-commutative) multiplication of string functionals, $*$, and an integration,$\int$. These obey the usual, with the exception of graded commutativity, axioms of multiplication, differentiation and integration and therefore, as the mathematician Connes has showed, allow for a natural generalization of gauge invariance. Witten argues that the form of the string Lagrangian is just the analogue, for this non-commutative geometry, of the Chern-Simons form $\mathcal{S} = \int[\Phi * Q\Phi + 2/3\Phi * \Phi * \Phi]$. The resulting equations of motion are simply $F = 0$, where F is the field strength, $F = Q\Phi + \Phi * \Phi$. What could be simpler !

This formulism has been developed quite far. It has been shown explicitly (by Giddings) to reproduce the four point scattering amplitude of the Veneziano model, to yield (using path integral techniques by Martinec and Giddings) the correct measure for the N-point function, and to yield the correct measure (by Giddings, Martinec and Witten) for the loop amplitudes. Jevitcki and I have worked out the explicit operator formulation of all of the ingredients of the theory, including the vertex operator, and have explicitly verified the symmetries of the action. Witten has generalized the approach to the open superstring. Much, however, remains to be done. So far this formulation has not been extended to the closed string, including the heterotic string; and to date no nontrivial use has been made of this formulism.

How Many String Theories Are There?

At the moment one can count four consistent string theories. There are the two forms (chiral and nonchiral) of the closed superstring, there is the SO_{32} open string theory of Green and Schwarz and there is the Heterotic string theory. It is now realized that the two manifestations of this theory (with gauge groups $E_8 x E_8$ or $Spin_{32}/Z_2$) are simply two vacuumn states of the same theory.

Do there exist more consistent theories than the known five? There have been some recent constructions of new models, but these too are probably just different (likely unstable) vacua of the heterotic string. Do there exist fewer, in the sense that some of the ones we know already are perhaps different manifestations (different vacua?) of the same theory? Although there have been speculation along these lines no convincing argument to this effect is known.

What is special about the heterotic string?

String Technology

This is not a question but a program of development of the techniques for performing calculations within string theory, including control of multiloop perturbation theory and the construction of manifestly covariant and supersymmetric methods of calculation. Much progress has been made in this area during the last year. Of special importance is the development of the superconformal field theory approach to the construction of superstring loop amplitudes, the construction of the fermionic vertex (by Friedan, Martinec and Shenker); and the application of sophisticated algebraic geometry to the evaluation of the integrands of multiloop amplitudes (by Belavin and Kniznik and by Manin). Enormous mathematical power is now being brought to bear in this area with remarkably beautiful and suggestive results.

What Is the Nature of String Perturbation Theory?

Does the perturbative expansion of the string theory converge ? As I remarked above, probably (and hopefully) not. If not, when does it give a reliable asymptotic expansion? How can one go beyond perturbation theory? Can one develop semi-classical methods (instantons) to treat the nonperturbative physics? Do the non-perturbative effects totally destroy the perturbative picture of the string vacuum? Here there are many questions and no answers.

String Phenomenology

Here there are many issues that remain to be resolved. They can all be included in the one question-can one construct a totally realistic model which agrees with observation and why is it picked out?

What Picks the Correct Vacuum?

During the last year much effort has been devoted to exploring the classical equations of motion of string theory. Mostly one wishes to find static solutions that might describe the background fields of stable quantum vacua. As described above, there are two approaches to the derivation of these equations. The first method, that of deducing the effective action from the string scattering amplitudes, has been pursued by Witten and myself; the second, based on demanding conformal invariance of the two dimensional σ model that describes the first quantized string in a nontrivial background space, has been discussed by Grisaru, Zannon, deVen, Pope and Freeman. Both groups agree that, to quartic order in the curvature (say of the internal manifold) , the effective Lagrangian (in the Type II superstring or in the Heterotic string with the spin connection embedded in the gauge group) differs from the Einsteinian Lagrangian, $\sqrt{g}R$, by terms quartic in the Riemann tensor R_{abcd}. These terms have an interesting effect on the issue as to whether the Calabi-Yao spaces are indeed solutions of the string equations to all orders. Stricly speaking they invalidate previously given proofs to that effect. However one can show that for every Calabi-Yao space there is another space close by (i.e. one whose metric differs by terms that are small, when the curvature of the internal manifold is small); and that can be calculated order by order in an expansion in the curvature. This requirement, to slightly shift the metric of the solution, has no effect on the resulting physics (the number of generations, the breaking or preservation of symmetries, etc.).

Other attempts have been made, notably by Witten, to extend the class of solutions to include ones where the spin connection is not embedded in the gauge group. This requires turning on torsion, and has the advantage of allowing a GUT group of SU_5 or SO_{10}, which solves many phenomenological problems. Unfortunately almost all such perturbative solutions have (σ -model) nonperturbative corrections (instantons), as shown by Dine, Seiberg and Witten. This does leave for consideration the very special stringy solutions known as orbifolds, which offer interesting possibilities that are just now being explored (by Harvery, Dixon and others).

Nevertheless there exist already an enormous number of acceptable vacuum states. What is the dynamics that chooses among all these possibilities? Why don't we live in ten dimensional flat space? How does the value of the dilaton field get fixed and thereby the dilaton acquire a mass? Does the vanishing of the cosmological constant survive the physical mechanism that lifts the vacuum degeneracy?

What Is the Nature of High Energy Physics?

By this I mean what does physics look like at energies well above the Planck mass scale? This is a question that is addressable, in principle for the first time, and might be of more than academic interest for cosmology, where is interested in the initial conditions of the universe. Traditionally this means pushing back to arbitrary early times where the temperature and density were arbitrarily high. At Planckian times, densities and temperatures string physics will be relevent. Does the string undergo a transition to a new phase at high temperatures and densities? Does string theory determine the intial state of the universe? Here no progress has been made in developing a string cosmology.

Another question of great academic interest is that of singularities. Einsteinian relativity has two severe problems. The first is the nonrenormalizabiltiy of the quantum theory. This problem appears to be solved by string theory. However there is also the problem of the ubiquitous singularities and the resulting incompleteness of ordinary relativity that already shows up at the classical level. It is perfectly possible that string theory will cure this problem. The above mentioned corrections (say, those arising from the quartic terms in Rabcd in the effective action) to Einsteinian relativity reopen the question of singularities. For example we do not know whether classical string theory contains black holes (Witten and I have speculated that the end product of collapsing matter might be nonsingular, dilaton radiation playing the role in the classical theory of Hawking radiation in the quantum theory). One could very well imagine that at very short distances the usual space-time description of physics breaks down, and that string theory avoids the ubiquitous singularities that plague ordinary general relativity already at the classical level, without having to resort to hopes of a quantum cure.

Is There a Measurable, Qualitatively Distinctive, Prediction of String Theory?

The Heterotic string theory can make, in principle, many postdictions (such as the calculation of mass ratios of quarks and leptons, Higgs masses, gauge couplings etc.) and it can make many new predictions (such as the masses of the various supersymmetric partners). These would be sufficient to establish the validity of the theory. However one could imagine conventional field theories coming up with similar pre or postdictions. It would be nice if there were a phenomenon which might be accessible at observable energies and is uniquely characteristic of string theory. Perhaps our experimental friends will find one for us.

DISCUSSIONS I

CHAIRMAN: D. Gross

Scientific Secretaries: C. Scott, D. Coccolino, S. Giddings and P. Vecsernyes

– *K. Meissner:*

In the light cone formulation of the heterotic strings, is the choice $x^+ \propto \tau$ globally well defined in the case of a scattering diagram with an arbitrary number of loops?

– *D. Gross:*

I am not aware of any problem with doing this in Minkowski space. Of course the curvature is singular at the joining points.

– *F. Quevedo:*

You have shown how states corresponding to gauges bosons of $E_8 \times E_8$ or $SPIN(32)/Z_2$ arise in the construction of the heterotic string. Is there a way to see that these groups are manifestly physical symmetries of the theory?

– *D. Gross:*

On physical grounds one expects that if there is a spin 1 massless particle in a theory, then for consistency there should be an associated local gauge symmetry. This can be explicitly demonstrated by constructing the conserved charges corresponding to the gauge symmetries. These are obtained by constructing the vertex operators for the emission of gauge bosons of zero momentum.

– *S. Giddings:*

What is the current thinking on the possibility that the heterotic or super strings come from the bosonic string through some kind of compactification mechanism?

– *D. Gross:*

Several people have claimed to show that one can obtain the heterotic or D = 10 superstring through compactification. In my opinion these attempts appear so far unrealistic. The problem is getting space–time fermionic operators of the correct form from the bosonic operators.

– *P. Vecsernyes:*

If you can describe changes of particle number in the first quantized formalism, why do we need a field theory for strings?

– *D. Gross:*

The first quantized formalism does not describe non–perturative phenomena. For example, in QCD the first quantized formalism corresponds to perturbation theory. If we use the perturbation series we do not see important non–perturbative phenomena such as confinement or chiral symmetry breaking. Also, I would expect that the perturbative expansion diverges and does not, by itself, define the theory.

– *P. Vecsernyes:*

If we choose another metric for the embedding space do we get the same result for the critical dimension?

– *D. Gross:*

Not necessarily. The requirement that D = 10 can be thought of as the first term in the equation of motion for the dilaton. Only in flat space does this equation of motion fix the dimension to be ten. One might imagine that as a consequence of background gauge fields and partial compactification, that D might be 9 or 11. In superstring theory, however, this is difficult to imagine since 10 is picked out on other grounds.

– *P. Vecsernyes:*

For an infinite dimensional algebra we know that there may be inequivalent representations of the canonical commutation relations. What about the case of strings?

– *D. Gross:*

Of course in ordinary quantum field theory we already have an infinite number of degrees of freedom, in string theory we just have a *"bigger"* infinity. The existence of many solutions do indeed give us a large number of unitarily inequivalent representations of the canonical commutation relations.

– *A. Shapere:*

Do you envisage a form of the heterotic string theory where the space–time structure would appear more naturally?

– *D. Gross:*

Conversely, I envisage that eventually, as we learn more about the non–perturbative aspects of string theory, space–time will appear as a more derived concept.

In other words the perturbative picture of strings propagating in space–time will become inadequate. Space–time will be a concept useful only at low energy.

– *P. Newman:*

Has anybody tried to modify the Nambu action?

– *D. Gross:*

Polyakov has written an action where a term involving the extrinsic curvature of the world sheet has been added. This was constructed in an attempt to describe QCD in the infrared. When studying fundamental strings, these terms would appear to destroy the delicate balance of two dimensional symmetries that are in turn responsible for the ten dimensional symmetries of string theory.

– *M. McGuigan:*

If one tries toroidal compactification to dimensions less than 10 do any new conditions arise on the gauge group?

– *D. Gross:*

Not necessarily, it depends on what you do. In the case of the heterotic strings the conditions arise because we have chiral left moving fermions. Narain has discovered some compactifications in which the gauge group is enlarged. In these cases new, stringent conditions are obtained.

– *S. Webb:*

One considers theories with different numbers of space–time dimensions; why don't we consider spacetimes with no time dimension or two or more time dimensions?

– *D. Gross:*

You are asking why the signature of space–time is $(+ - - - \cdots)$ instead of $(- - - \cdots)$ or $(+ + - - - - \cdots)$. This is one of the few questions which it seems difficult for string theory to answer.

It is difficult to interpret a theory of physics with no time dimension and there can be problems with causality if there are two times. Sakharov has discussed the

Kaluza–Klein compactification of theories with more than one time; the extra time dimension must be a small circle so we don't see it. It's not clear if one can escape causality problems, and the interpretation of a theory with two times is not clear.

– *Z. Bern:*

Can the tachyonic ground state of the string be eliminated by some kind of shift in the vacuum (as in conventional field theory)?

– *D. Gross:*

The existence of the tachyonic ground state of the bosonic string could mean that there exists another stable ground state (presumably a curved manifold) or that the theory has no ground state. To explore this possibility, to *"shift"* to a new vacuum, one requires methods which go beyond first quantization, which only allows one to construct perturbation theory about a given vacuum state.

DISCUSSION II

– *G. D'Ambrosio:*

Do you think that superstrings can solve the hierarchy problem?

– *D. Gross:*

Hopefully. The hierarchy problem is supposed to be solved by supersymmetry. Of course we have to break SUSY and this can only be done non–perturbatively. It is an open question in superstring theory how this is done, and whether $N = 1$ SUSY can survive down low energies.

– *H. Livne:*

What type of heterotic string theories can you obtain if you compactify more than 16 coordinates? Are they different theories as regards to spectrum?

– *D. Gross:*

Today I considered particular compactifications of $16_L + 6_{L,R}$ dimensions, but there are alternative schemes. For example there are those due to Narain which give a larger gauge group than $E_8 \times E_8$. These alternative compactifications are also solutions but appear not to be phenomenological interest, because we want a smaller gauge group, chiral fermions and $N = 1$ SUSY.

Orbifolds–toroidal compactifications with twisted boundary conditions provide a different type of compactification that is potentially realistic.

– *P. Fisher:*

You introduce interactions by putting handles in the world sheet of the string and summing over all possible topologies. Why can't you carry out a similar procedure in ordinary quantum field theory?

– *D. Gross:*

In a sense you do. The difference is that there is nothing continuous about a line breaking up. This is a genuinely singular operation. In a 2–dimensional surface there is no singularity. In ordinary field theory you have freedom to choose the form of the interaction, whereas in string theory you do not.

– *J. Quackennbush:*

In the heterotic string theory you compactify 16 dimensions of the bosonic sector. What causes this compactification, and is there any problem with inconsistency?

– *D. Gross:*

Whether or not you compactify the 16 dimensions has no relevance to the appearance of the conformal anomaly. This is a local question, and a torus is locally indistinguishable from flat space.

The two particular tori used are, as we have seen, determined by the demands of chirality, two–dimensional reparametrization invariance, and modular invariance. If we wish, we can consider these conditions as a sort of "equation of motion" fixing the compactification.

– *A. Pasquinucci:*

The compactifications on Calabi–Yan manifolds, with vanishing cosmological constant, appear to come from D = 10 supergravity. Why do we need superstrings?

– *D. Gross:*

Anomaly free D=10 supergravity is unique. One can argue that since it is the low energy limit of the superstring, it is (almost) consistent. In fact D=10 supergravity is not quite consistent, since it is presumably non–renormalizable. The string compactifications, with zero cosmological constant, that have been studied are in fact solutions of the equations of motion of D=10 supergravity. But they also solve the equations that contain the higher order terms that come from string theory.

– *A. Pasquinucci:*

In 10–dimensions we have $E_8 \times E_8$ as a gauge group. On compactifying to 4–dimensions one of the E_8's breaks to give an E_6 grand unification group. What about the other E_8?

– *D. Gross:*

The emergence of matter in the first E_8 is dictated by the dynamics. The other E_8 doesn't have to be broken. You can find solutions where it is broken, but not by the same mechanism as the first E_8. Because of this there is no matter with quantum numbers of the second E_8. If the second E_8 is not broken it contains zero mass gluons whose interactions become strong only at the Planck mass. This sector then contains only massive glueballs. The two E_8 sectors couple only via gravity.

The second E_8 can, however, play an important role. For example it has been suggested that SUSY is broken in this "hidden sector" by the formation of a gluiono condensate. This SUSY breaking would then feed into our sector by gravitational interactions.

– *M. Carter:*

When I think about whether a Kaluza–Klein manifold is stable, I think about calculating an effective potential and seeing if it has an imaginary part. If it has an imaginary part, the space has runaway expansion or contraction, how does this work in string theories?

– *D. Gross:*

In Kaluza–Klein theory, there is no easy way to calculate an effective potential. In order to examine classical stability we look at the excitations. If there are tachyons, which describe runaway modes, this is an indication of instability. In string theory, we have an embarassment of riches. We have 10^4 or more stable vacua parametrized by continuous parameters. The "potential" has many flat directions, and because of supersymmetry, classical stability implies quantum stability to all orders in perturbation theory. The hope is that there is a non–perturbative quantum mechanical mechanism that destabilizes, for example, 10–dimensional Minkowski space so that it decays to our world.

– *S. Giddings:*

You mentioned that it is possible to make local field redefinitions. How does one then say what one means by an individual field such as the metric or the graviton?

– *D. Gross:*

There is no direct meaning in quantum gravity to the metric. Only observable correlation functions, describing scattering of gravitons, have meaning. The fields are like a set of coordinates, and we are free to change coordinates, for example, we could take

$$g_{\mu\nu} \to g_{\mu\nu} + R_{\mu\nu} + \text{(higher order terms)}$$

$$g_{\mu\nu} \to g_{\mu\nu} + R_{\mu\nu} + \text{(terms involving other fields)}$$

without modyfing the physics.

– *M. McGuigan:*

Are there singularities and black holes in string theory?

– *D. Gross:*

Nobody knows. But I like to speculate that string theories will solve the problems of singularities. If we could obtain the equations of motion of string theories then we could investigate singularities in gravity. Einstein's theory of gravity is incomplete on two grounds. Firstly, it does not have a consistent quantization and secondly, there exist singularities. Quantum mechanics, instead of "smoothing-out" the singularities actually appears to make things worse. We can hope that string theory is complete – the (singular) Schwarzchild metric is not a solution of string theory. The existence of the dilaton mode of the string might provide a mechanism for the radiation of energy of a collapsing mass.

ANOMALIES, STRINGS AND ALGEBRAIC GEOMETRY

Luis Alvarez-Gaumé

Theoretical Physics Division
CERN
1211 Geneva 23, Switzerland

1. - INTRODUCTION

In these lectures we would like to review some of the applications of global methods in geometry and topology to the quantum theory of fields and strings. The use of the theory of characteristic classes, index theory, algebraic geometry and differential topology is becoming common place in recent theoretical physics. It seems likely that these areas of mathematics will soon become part of the mathematical background of theoretical physicists, in the same way that Riemannian geometry, group theory, and complex analysis became part of our mathematical arsenal. These new techniques provide also rather powerful tools to obtain useful qualitative (and sometimes quantitative) information in quantum field theories with gauge and/or gravitational interactions, and in string theory.

In the first part of these lectures, we shall emphasize the use of global methods in the study of local and global anomalies in field theory; in the second part we shall explain some of the applications in string perturbation theory. Thus in Section 2 we will give a rather quick review of local and global anomalies; and in Section 3 we will present some recent results on multi-loop string diagrams and bosonization techniques on Riemann surfaces with many handles.

2. - ANOMALIES IN QUANTUM FIELD THEORY[1]

We can divide the anomalies appearing in quantum field theory into two large categories: local and global. Local anomalies are those accessible by purely perturbative methods. After we have checked that a theory is free of local anomalies, we are still left with the problem of checking that it is also free of global anomalies. The latter measure the change of the effective action for a set of fields in the presence of gauge and gravitational backgrounds under gauge transformations or diffeomorphisms not connected with the identity. The study of these anomalies has been carried out by Witten[2], and we will review some of the basic results. In string theory, invariance under the group of disconnected diffeomorphisms on a Riemann surface plays a very important role in determining the consistency and the spectrum of the theory. Fortunately, the same methods developed in Ref. 2 can be applied to this case[3] to show that the set of fields needed in all known string theories to cancel the conformal anomalies in the string worldsheet guarantee that the theory can be written in a modular invariant way.

A) Local Anomalies

In its original form[4] one considers a massless fermion triangle diagram with one axial current and two vector currents. Regulating the graph with the requirements of current conservation and Bose symmetry in the vector channels, one finds that the axial current is not conserved. This fact has had important phenomenological implications [decay of the π° [4], resolution of the U(1)-problem in QCD[5]]. By analogy, if instead of considering two vector currents in the triangle graph, one inserts two energy momentum tensors corresponding to the emission of two gravitons[6], the same conclusion follows. The axial current is not conserved in the presence of external gravitational fields. If we now consider V-A currents coupled to gauge fields in the vertices of the triangle graph, the diagrammatic computation is essentially unchanged, but now the presence of the anomaly would imply the loss of gauge invariance and renormalizability. The cancellation of anomalies in this case provides some constraints on the fermion spectrum in chiral gauge theories. The best known example is probably the cancellation of anomalies between quarks and leptons in the standard SU(3)×SU(2)×U(1) theory of strong weak and electromagnetic interactions. The representation content of a family of quarks and leptons is, in terms of left-handed Weyl fermions given by

$$(3,2)_{1/6} \oplus (\bar{3},1)_{-2/3} \oplus (1,2)_{-1/2} \oplus (\bar{3},1)_{1/3} \oplus (1,1)_{1} \qquad (1)$$

where the first entry indicates the SU(3) representation, the second one gives the SU(2) representation, and the third one is the value of the weak hypercharge. To check that the anomalies cancel, one has to make sure that the following condition is satisfied[7]:

$$\sum_L \mathrm{tr}\, T_L^a \{T_L^b, T_L^c\} - \sum_R \mathrm{tr}\, T_R^a \{T_R^b, T_R^c\} = 0 \qquad (2)$$

where the T_L's (resp. T_R's) are the generators of the gauge group in the representation of the left- (resp. right-) handed fermions. The remarkable cancellation of anomalies between quark and lepton (which the reader is encouraged to check) lends substantial support to the family structure in the standard model. If one considers the computation in Ref. 6 for a V-A current, one finds that the triangle graph with one gauge field and two gravitons is proportional to the trace of the charge associated to the gauge field. Since the trace of a generator in a semi-simple group is always zero, this possible anomaly would only show up in gauge groups containing U(1) factors. It is easy to check that in the standard model the trace of the weak hypercharge generator vanishes in each family. The easiest way to show it is perhaps that the standard model can be embedded in an SU(5) GUT theory without changing the number of fermions[8].

For theories in more than four dimensions, one finds a potentially anomalous diagram analogous to the triangle graph in any even number of dimensions. In d = 2n dimensions, the first potentially dangerous graph is a polygon with n+1 vertices. The anomaly can be evaluated by diagrammatic techniques as in four dimensions[9], and the group theory factor in front of the kinematical form of the anomaly is

$$\sum_L \mathrm{Str}\, T_L^{a_1} \dots T_L^{a_{n+1}} - \sum_R \mathrm{Str}\, T_R^{a_1} \dots T_R^{a_{n+1}} = 0 \qquad (3)$$

Str means the symmetrized trace which follows from the Bose symmetry and we use the same notation as in (2). A more concise form of writing (3) is to consider an arbitrary element of the Lie algebra of the gauge group X. Then the cancellation of (3) is equivalent to the cancellation of

$$\sum_L \mathrm{Tr}\, X_L^{n+1} - \sum_R \mathrm{Tr}\, X_R^{n+1} = 0 \tag{4}$$

An important difference between d = 4k and d = 4k+2 can already be noticed in (4). In 4k dimensions we can have purely left-handed anomaly free theories. This is because in this case we have to compute $\mathrm{tr}X^{2k+1}$ (an odd power). Thus, for example, if the representation is real or pseudoreal, we know that $\mathrm{tr}X^{2K+1} = 0$, for in this case ${}^tX = -SXS^{-1}$, and we can conclude in four dimensions that Sp(n), SO(2n+1), SO(4n), G_2, F_4, E_7 and E_8 are free of anomalies in any 4k-dimensions. Anomalies would only appear for groups SO(4n+2), U(n) and E_6. Instead, if we consider d = 4k+2, then the anomaly for a given representation is proportional to $\mathrm{tr}X^{2K+2}$. Since X is Hermitian, this trace is always positive and it is not possible to have a purely left-handed anomaly free theory in d = 4k+2. The reason for this difference (which is also at the origin of gravitational anomalies) is the interplay between chirality and charge conjugation. In 4k+2 (resp. 4k) dimensions, charge conjugation commutes (resp. anticommutes) with Γ (the higher dimensional analogue of γ_5). One should recall that, for example in d = 2, both particles and antiparticles have the same helicity, while in d = 4 they have opposite helicity. In particular, the purely gravitational coupling of a Weyl fermion in d = 4k+2 is genuinely chiral, and the theory may be afflicted with anomalies in the energy-momentum tensor[10]. If at the same time the fermions are also coupled to external gauge fields, we can in general have three types of anomalies: purely gravitational, purely gauge, or mixed anomalies depending on whether we have only gravitons or only gauge fields or both on the external legs of the (n+1) polygon. In particular, in d = 10, the first potentially anomalous graph is the hexagon, and in the external lines we can have: (i) six gravitons, (ii) six gauge fields; (iii) two gravitons and four gauge fields, or (iv) four gravitons and two gauge fields. The constraints imposed on a higher-dimensional chiral theory by the cancellation of anomalies are highly restrictive, and for a while the only supersymmetric theory known to have a non-trivial cancellation of anomalies was the chiral N = 2 supergravity theory in ten dimensions[10] (or its compactifications). The whole picture changed drastically with the discovery of a new anomaly cancellation mechanism by Green and Schwarz[11]. They showed that in d = 10, the N = 1 supergravity multiplet coupled to the N = 1 super-Yang-Mills theory is anomaly free provided that the gauge group is either SO(32) or $E_8 \times E_8$. Soon after the discovery of the anomaly cancellation, the heterotic string was constructed[12] (see Prof. Gross' lectures for more details), and the string Big Bang begun. Before explaining in more detail the Green-Schwarz mechanism, we would like to review the general prescription for the computation of anomalies in any number of dimensions.

The general prescription for the computation of local anomalies can be given in a very elegant way in terms of the Atiyah-Singer index theorem[13]. For the case of global gauge singlet currents, such as the axial current in QCD, it was shown some years ago[14] that the tensor structure and normalization of the anomaly follow from the classical index theorem applied to the Dirac operator in the presence of gauge and gravitational fields. The topological interpretation of the anomaly in gauge currents requires more powerful machinery. In this case, one has to appeal to the more general form of the Atiyah-Singer index theorem which applies to families of elliptic operators. In order to make things more explicit, we consider in detail the case of a Weyl fermion in 2n dimensions in Euclidean space coupled to an arbitrary external gauge field. The starting point of the analysis is always to consider the regularized fermionic effective action

$$\exp(-\Gamma[A]) = \int d\lambda d\bar{\lambda} \exp - \int \bar{\lambda} D(A) \lambda \tag{5}$$

where D(A) is the Weyl operator, and it is understood that some regulator has been chosen for the functional integral. We take A to be an anti-Hermitian one-form valued in the representation of the Lie algebra of the gauge group carried by the fermions. An infinitesimal gauge transformation of A is given by

$$\delta A = dv + [A, v] = Dv \tag{6}$$

and v is a scalar valued in the same representation as A. Gauge invariance means that up to trivial local counterterms, the effective action does not change under gauge transformation

$$\Gamma[A+Dv] - \Gamma[A] = -\int dx \, \mathrm{Tr} \, v \, D \, \delta\Gamma[A]/\delta A$$

$$= \int dx \, \mathrm{Tr} \, v \, D_A \langle j(A) \rangle \tag{7}$$

where j(A) is the fermion current, and the angular brackets represent as usual the expectation value with respect to the measure defined by (5). As a consequence of Eq. (7), we learn that gauge invariance is equivalent to current conservation. From the topological point of view, $\exp(-\Gamma[A])$ defines a line bundle over the space of gauge configurations. In this language, the anomaly is simply expressed by saying that the line bundle is not flat; in other words the connection which determines the parallel transport of the effective action over the space of gauge fields is not flat and has a non-trivial curvature. In analogy with the monopole theory, we can test for the non-triviality of this line bundle by restricting it to topologically non-trivial two surfaces in the space of gauge connections modulo gauge transformations[15]. Here is where the index theorem for families of elliptic operators[16] becomes useful, because it provides a local representative of the first Chern class of this line bundle (its local curvature). We can now briefly summarize the results obtained over the last three years concerning the form of this curvature (for further details and references, see Ref. 1). Let F be the gauge field strength two-form $F = dA+A^2$, and let R be the curvature two-form associated to the external gravitational field. If we work in orthonormal frames, R is a two-form valued on the Lie algebra of the frame rotation group, SO(d): The index for the Dirac operator coupled to A and the spin connection ω, $R = d\omega+\omega^2$, is obtained by the following procedure. Since R is an antisymmetric matrix of two-forms, we can formally skew diagonalize R so that it looks like

$$\begin{bmatrix} 0 & x_1 & & & \\ -x_1 & 0 & & & \\ & & \ddots & & \\ & & & 0 & x_n \\ & & & -x_n & 0 \end{bmatrix} \tag{8}$$

and construct the formal polynomial

$$A(M) = \prod_i (x_i/4\pi) \Big/ \sinh(x_i/4\pi) \tag{9}$$

known as the Dirac genus of the manifold M where the fermion lives. Notice that even though A(M) is an infinite series in the curvature eigenvalues, the series actually terminates because the x's are two-forms. Then the index for the Dirac operator is obtained by expanding the polynomial

$$\mathrm{Ind}\, D(\omega, A) = A(M)\, \mathrm{Tr}\, \exp F/2\pi i \tag{10}$$

until we find a term whose degree is equal to the dimension of M, and then integrating over M. This local polynomial gives the anomaly for the global axial current. Notice that due to the symmetries of A(M), all the terms in its Taylor series expansion can be re-expressed in terms of R as combinations of monomials of the form Tr R^n. In four dimensions, (10) takes a familiar form

$$(2\pi)^2\, \mathrm{Ind}\, D(\omega, A) = -\mathrm{Tr} F^2/2 + r\, \mathrm{Tr} R^2/48 \tag{11}$$

where r is the number of fermions. We can easily recognize the ABJ anomaly in (11) when se set R = 0. (We are using a notation where the wedge product between forms is left implicit.) The anomaly for gauge currents can also be stated in very similar terms. A fundamental property of the differential forms entering the Taylor series expansion in (7) is that they are closed, therefore they can be expressed locally as total derivatives. For example, if we consider the 2N+2-form Tr F^{N+1}, one can write

$$\begin{aligned} \mathrm{Tr}\, F^{N+1} &= (N+1)\, d \int dt\, \mathrm{Tr}\, A\, (t dA + t^2 A^2)^N \\ &= d\, Q^0_{2N+1}(A, F) \end{aligned} \tag{12}$$

where Q is known as the Chern-Simons form[17]. If we make a gauge transformation of Q^0_{2N+1}, it can be shown that[18]

$$\delta_v Q^0_{2N+1} = d\, Q^1_{2N}(v, A) \tag{13}$$

and that the change in the effective action under a gauge transformation is given by Q^1_{2n} if the dimension of space time is 2n. The prescription to calculate the anomaly is then to first compute the index polynomial for the operator involved, depending on the spin fields we work with (spin ½ or spin 3/2 or self-dual tensor fields...) and the couplings involved. Next expand this polynomial to obtain the form of dimension 2n+2. Using (12) and (13) but now applied to the term in the index polynomial of degree 2n+2, the variation of the effective action is finally given by

$$\delta_v \Gamma[A] = \int Q^1_{2n}(v, A) \tag{14}$$

in four dimensions for example we have

$$Q^1_4(v, A) = \frac{1}{24\pi^2}\, v\, d\, (A dA + \tfrac{1}{2} A^3) \tag{15}$$

and similar expressions can be found for higher dimensions (see Ref. 1 for more details). In particular, if we concentrate in d = 10, and consider the contribution to the anomaly due to all the fields of the N = 1 supergravity and super Yang-Mills multiplets, the result can be succintly summarized by the following 12-form

$$\begin{aligned} I_{12} = &-\frac{1}{15}\mathrm{Tr} F^6 + \frac{1}{24}\mathrm{Tr} F^4\, \mathrm{Tr} R^2 - \\ &- \frac{\mathrm{Tr} F^2}{960}\left[5(\mathrm{Tr} R^2)^2 + 4\, \mathrm{Tr} R^4\right] + \\ &+ \frac{N-496}{7560}\mathrm{Tr} R^6 + \left[\frac{N-496}{5760} + \frac{1}{8}\right]\mathrm{Tr} R^4\, \mathrm{Tr} R^2 \\ &+ \left[\frac{N-496}{13824} + \frac{1}{32}\right](\mathrm{Tr} R^2)^3 \end{aligned} \tag{16}$$

in the standard way if cancelling anomalies one would require that (16) vanishes identically when the contribution from all fields has been included. It is easy to see that (16) cannot possibly vanish for any gauge group. What Green and Schwarz noticed is that one can use the antisymmetric tensor field which appears in the massless sector of the string in order to construct counterterms whose variation will cancel the gauge variation of the effective action represented by (16). The new prescription for anomaly cancellation can be summarized by saying that if (16) factorizes in the form

$$I_{12}(F,R) = (\mathrm{Tr}\,R^2 - k\,\mathrm{Tr}\,F^2)\,X_8 \qquad (17)$$

where X_8 is an invariant form, then such counterterms always exists. In the case of (16), (17) is satisfied only if the gauge group is $E_8 \times E_8$ or SO(32) [and also $U(1)^{496}$]. The factorization condition requires the gauge group to obey the equation

$$\mathrm{Tr}\,F^6 = \frac{1}{48}\mathrm{Tr}\,F^2\,\mathrm{Tr}\,F^4 - \frac{1}{14400}(\mathrm{Tr}\,F^2)^3 \qquad (18)$$

as well as $k = 1/30$.

It is worth pointing out that the prescription (12)-(13) is part of the descent equations of Stora and Zumino[19], and can be obtained via purely algebraic methods using the BRS algebra, or by purely topological methods using the index theorem for families of operators.

Other interesting aspects of the recent developments in anomalies which we do not have time to review are the following.

a) The Hamiltonian interpretation of anomalies[20]. If one starts from a purely Hamiltonian formulation of a theory of Weyl fermions coupled to an external gauge field, the anomaly is found as a Schwinger term in the commutator of the gauge currents. The Schwinger term represents a two-cocycle in the algebra of infinitesimal gauge transformations, and it measures the obstruction to the possibility of constructing a fermionic Fock vacuum satisfying the Gauss law.

b) Non-linear σ-model anomalies[21]. These are anomalies appearing in low energy effective Lagrangians where the symmetries are realized non-linearly. Apart from the many applications in constraining the chiral structure of low energy effective theories, the σ-model anomalies provide a novel way of understanding the 't Hooft anomaly matching conditions[22].

c) The equivalence of local Lorentz and gravitational anomalies. When considering the coupling of fermions to gravity, we have to introduce local Lorentz frames. The reason is that the frame rotation group does admit spinor representations, whereas the general linear group does not. Thus a theory of fermions coupled to gravity possesses two types of invariances: general co-ordinate transformations and local frame rotations. If the fermions are chiral, one would naively expect two different kinds of anomalies. Happily this is not the case. In Ref. 22, Bardeen and Zumino showed that they are equivalent by explicitly exhibiting the local counterterm interpolating between them.

d) Anomalies in odd dimensions[23,24]. These "parity" anomalies appear for certain representations of fermions in odd dimensions [for example, an odd number of SU(2) doublets in three dimensions[24]]. The fermionic effective action cannot be regulated in such a way as to preserve both gauge invariance and parity. The parity violating part of the effective action is given by the Chern-Simons form (12) if we work in flat space, and by the Atiyah, Patodi Singer η-invariant (see part B) if we work in curved space[24]. These anomalies have a number of applications in problems related to fermion fractionalization (see Ref. 25 for details).

These remarks conclude our quick tour of local anomalies.

B) Global Anomalies

Even when we have cancelled the local anomalous variation of the effective action under infinitesimal gauge transformations or diffeomorphisms, we still have to make sure that the theory is also invariant under those transformations not connected to the identity. We can illustrate this problem by working again with a gauge theory with fermions in Euclidean space-time compactified to a sphere S^{2n}. The group of gauge transformations is the set of maps from S^{2n} into the gauge group G. Since we have compactified Euclidean space by adding the point at infinity, we will require that the gauge transformations be equal to the identity in the north pole. These maps fall into homotopy classes given by the elements of the homotopy group $\pi_0(S^{2n}\to G)$. The elements in the trivial class of π_0 are those relevant in the computation of local anomalies, and the cancellation of local anomalies is equivalent to the statement that $\Gamma[A]$ is invariant under those gauge transformations in the identity of π_0. A more difficult, but an equally relevant question, is whether $\Gamma[A]$ will be invariant under the other elements of $\pi_0(S^{2n}\to G) = \pi_{2n}(G)$. To answer this question we cannot consider a one-parameter family of gauge transformations interpolating between the identity and a non-trivial element of $\pi_{2n}(G)$. However, we can take advantage of the fact that the space of gauge connection on the sphere for a fixed topological class is a contractible space. The deformation retract is given as follows. Given any two gauge connections A and A' in the same topological class we can construct the one-parameter family of gauge fields $A(t) = A+t\,(A'-A)$ which interpolates between A and A' as t changes from 0 to 1. Using this property we can interpolate continuously between A and A^g, where g is a gauge transformation in a non-trivial class of $\pi_{2n}(G)$. This is the procedure used by Witten[2] to evaluate global anomalies. Thus if we want to compare $\Gamma[A^g]$ with $\Gamma[A]$ we consider instead a one-parameter family of effective actions $\Gamma[A(t)]$. The conclusion[2] is that

$$\frac{e^{-\Gamma[A^g]}}{e^{-\Gamma[A]}} = e^{i\pi\eta(0)} \tag{19}$$

where η is computed as follows. Since we have constructed a one-parameter family of gauge fields, we can construct the mapping cylinder $(S^{2n}\times[0,1])_g$ which means that on the cylinder $S^{2n}\times[0,1]$, the top and the bottom are identified after we twist by the gauge transformation g. The mapping cylinder is a compact odd dimensional manifold, and we can consider the 2n+1 dimensional Dirac operator coupled to the gauge field A(t). In odd dimensions, the index of an elliptic operator vanishes (there is no chirality in odd dimensions), but there is another object which plays a central role in the index theorem for manifolds with boundaries[26]: the η-invariant of Atiyah Patodi and Singer. The η-invariant for a self-adjoint operator H on a compact manifold is defined by

$$\eta(s) = \sum_\lambda \operatorname{sgn}(\lambda)\,|\lambda|^{-s} \tag{20}$$

for Re(s) large enough, the series (20) converges (the sum is over the eigenvalues of H), and it can be analytically continued to the rest of the complex s-plane. For a large class of operators (including the Dirac operator), $\eta(0)$ is well defined, and roughly speaking, it measures the number of positive minus the number of negative eigenvalues of the operator involved. It is often called the spectral asymmetry of H. Witten's result implies that global gauge anomalies are determined by the spectral asymmetry of the Dirac operator on the mapping cylinder. When the local anomalies cancel between several representations, the exponent on the right-hand side of (19) becomes

$$i\pi\left(\sum_L \eta_L(0) - \sum_R \eta_R(0)\right) \tag{21}$$

where the subscript L(R) stands for the left- (right-) handed representations of fermions. Using the index theorem for manifolds with boundary, it is easy to show that (21) is a topological invariant depending only on the homotopy class of g and not on the particular gauge field we choose or even what metric we have on the sphere. To see this explicitly, we write down the formula for the index of the Dirac operator in a manifold M with boundary $\partial M = B$[26]:

$$\mathrm{ind}\, D = \int_M \mathrm{Ind}D(A,\omega) - (\eta(0) + h)/2 \tag{22}$$

where $\mathrm{Ind}D(A,\omega)$ is the index polynomial constructed in (10), η is the η-invariant calculated for the restriction of the Dirac operator to the boundary B, and h is the dimension of the kernel of the boundary Dirac operator. To show that (21) is a topologically invariant object when the local anomalies cancel, consider a deformation of the configuration that we chose for the mapping cylinder (for simplicity, we assume that h is equal to zero, although a more careful argument can be carried through where this is not the case). If we call I the interval where the parameter that defines the deformation varies, we can construct a 2n+2 dimensional manifold $D = (S^{2n}\times[0,1])_g \times I$ whose boundary is given by the original mapping cylinder and its deformation. Using now (22), we find

$$\sum_L \Delta\eta_L(0) - \sum_R \Delta\eta_R(0) = -2\sum_L \mathrm{ind}\, D_L + 2\sum_R \mathrm{ind}\, D_R$$
$$+ \int_D \left(\sum_L \mathrm{Ind}D(A,\omega) - \sum_R \mathrm{Ind}D(A,\omega)\right) (\mathrm{mod}\ \mathbb{Z}) \tag{23}$$

where $\Delta\eta$ represents the change in η in the interpolation between the initial and final mapping cylinders. Since the anomaly is given by the index theorem in two higher dimensions, the last two terms in (23) cancel if the combination of fermions chosen is free from local anomalies. Since the index is an integer, we conclude that (22) is a topological invariant which only depends on the topological class of the gauge transformation g. When the field considered is not simply a Weyl fermion, Eq. (16) needs some changes, but the basic idea is the same. When this analysis is applied to the chiral N = 2 ten-dimensional supergravity theory or to the $E_8\times E_8$ or SO(32) super-Yang-Mills theory coupled to N = 1 supergravity in ten dimensions, the result is that these theories are free of global gauge and gravitational anomalies[2].

In string theory, invariance under global diffeomorphisms on the string world sheet is very important in the determination of the string spectrum. In the NSR (Neveu-Schwarz-Ramond) formulation of the superstring[27], and of the heterotic string[12], the supersymmetric spectrum is obtained after one performs the GSO[28] projection. This projection can be interpreted geometrically as a sum over all the spin structures of the left- and right-handed fermions involved in the first quantized formulation of the theory on the Riemann surface traced out by the string. The relevance of the global anomalies in this case is that spin structures are not invariant under arbitrary non-trivial diffeomorphisms on the Riemann surface. Thus the absence of global anomalies in the string world sheet imposes severe constraints on the allowed spectrum of consistent string theories. For the known string theories, Witten[29] used his general analysis of global anomalies to show that once the local conformal anomalies are cancelled, there are no global anomalies with respect to diffeomorphisms which do not change the spin structure of the fermions. As a consequence, one can always choose signs in the sum over the spin structures (the GSO projection) consistent with invariance under global diffeomorphisms on the Riemann surface (modular invariance). The proof[3] of invariance under global diffeomorphisms in two dimensions is much simpler than in the ten-dimensional case. It follows from some general theorems concerning three-manifolds. For illustrative purposes, let us look at the computation of (19) in the case of a spin-$\frac{1}{2}$ Weyl fermion on a compact Riemann surface Σ of genus g (g is the number of handles). If $\mathrm{Diff}^+\Sigma$ stands for all diffeomorphisms of Σ preserving orientation, and $\mathrm{Diff}^+{}_0\Sigma$ is the normal subgroup of $\mathrm{Diff}^+\Sigma$ of those diffeomorphisms connected to the identity, the group of non-trivial diffeomorphisms of Σ is defined as $\Omega(\Sigma) = \mathrm{Diff}^+\Sigma/\mathrm{Diff}^+{}_0\Sigma$ [in the mathematic literature, $\Omega(\Sigma)$ is known as the mapping class group. Following the physics literature we will call $\Omega(\Sigma)$ the modular group of Σ]. There is a good deal of information about $\Omega(\Sigma)$[30]. For instance, it is known that one can always find a representative of each class in $\Omega(\Sigma)$ in terms of Dehn twists: given a closed curve γ on Σ, we can cut Σ along γ. Keeping one of the lips of the cut fixed, we twist the other one by 2π and glue them back together. Thus the mapping cylinder $(\Sigma\times I)_\phi$ is now constructed by gluing the top and bottom of $\Sigma\times I$ up to a Dehn twist. If the diffeomorphism preserves the spin structure chosen on Σ, then we get a spin 3-manifold bounds $(\Sigma\times I)_\phi$, and from (19), we have to compute the η-invariant to find any possible anomalies. The first useful mathematical result in this direciton is the fact that any spin three-manifold bounds a spin 4-manifold[31]. Using the Atiyah-Patodi-Singer index theorem[26], Eq. (22), in four dimensions we can write

$$\frac{\eta(0)}{2} = \frac{\sigma}{8} - \int_{B,\ \partial B = (\Sigma\times I)\phi} \hat{A}(R) \tag{24}$$

where σ is the signature of the 4-manifold B. The second mathematical result needed is a theorem of Rohlin that the signature of any spin 4-manifold with boundary is always equal to 0 (mod 8). The computations in d = 10 are much more subtle[2], but the main idea is the same: one always looks for a d+2-dimensional manifold whose topological invariants are known, and such that its boundary is the mapping cylinder. For more details, see Ref. 2.

Once modular invariance is proved, the understanding of the various ways of realizing modular invariance in the string world sheet is the reason behind the construction of the tachyon-free non-supersymmetric heterotic string $O(16)\times O(16)$[32]. In all known examples of string theories (even those with tachyons) with gauge groups other than $E_8\times E_8$ or SO(32), one always finds that once the theory is free of global anomalies in the string world sheet, it is also free of both local and global anomalies in ten dimensions. Modular invariance is one of the central elements in the formulation of string theories. Unfortunately we do not yet have a good understanding of this intimate interplay between the discrete group of global diffeomorphisms in two dimensions and the continuous group of gauge and co-ordinate transformations in d = 10. A first step in this direction is provided by the work of Ref. 33 where, for the first time, one can find a direct link between modular invariance and absence of local anomalies in the critical dimension.

2. - TOPOLOGICAL METHODS IN STRING PERTURBATION THEORY

In the first quantized formulation of string perturbation theory[27], one considers a string propagating in a fixed external background space-time. If we restrict our attention to closed strings, the loop expansion for on-shell amplitudes is an expansion in the number of handles of the two-dimensional surface describing the history of the string. After taking proper account of gauge fixing, one finds that in a given order of perturbation theory the Polyakov measure[34] requires that we sum over all inequivalent complex structures on a given surface of fixed genus. The space of all such complex structures is known as the moduli space of the surface, and it has 6g-6 dimensions. Even though the topology of this space is rather complicated, it has a fundamental property, namely it has the structure of a complex manifold (more precisely, a complex algebraic variety) whose boundary is made out of Riemann surfaces with nodes. A very particular property of moduli space is that even though it is a non-compact complex space, any globally defined holomorphic function is a constant[35]. When one considers string perturbation theory in the critical dimension, it can be shown[36] that the integrand of the closed string partition function factorizes into a holomorphic function times an anti-holomorphic function (or sums of factors with this general form if one wants to consider non-supersymmetric strings). Thus one can rephrase many questions of string perturbation theory (and perhaps even non-perturbative properties of the theory if one follows the infinite genus approach of Friedan and Shenker (see Friedan lectures in this volume) in terms of complex algebraic geometry. In the string path integral one integrates the partition and correlation functions of two dimensional field theories over the moduli space of conformally inequivalent Riemann surfaces. On a Riemann surface any field can be described in terms of q-differentials: $t = t(z,z^*)dz^q$, $q = 0, \pm\frac{1}{2}, \pm 1, \ldots$; when q is half integral, one has to be careful in choosing consistent branches for the square root (spin structures, see for instance 37). The kinetic term giving the dynamics of the field t always involves in the cases of interest in string theory, the Cauchy-Riemann operator $\bar{\partial}$ for the given complex structure. Thus the evaluation of string partition and correlation functions is reduced to the computation of functional determinants for $\bar{\partial}$-operators acting on q-differentials as well as their Green's functions. A very powerful procedure to carry out these calculations is to use bosonization as was shown by Friedan, Martinec and Shenker[38] when the fields live on a simply connected surface (string-tree level). The generalization of the bosonization prescription to surfaces of higher genus presents several difficulties, in particular the lack of a canonical quantization procedure for fields on compact surfaces of genus $g > 1$. The approach taken in 39 is to carry out the proof of bosonization in terms of Euclidean path integrals using algebraic geometry. The main ingredients in the mathematical proof of the results presented below for bosonization in surfaces of high genus are: (i) a theorem due to Quillen[40] which describes the determinant of a holomorphic family of wave operators, and (ii) the application of Faltings' work on Arakelov geometry[41], in particular, Faltings' inductive construction of determinants for the $\bar{\partial}$-operator acting on sections of any holomorphic line bundle over the Riemann surface in terms of spin-$\frac{1}{2}$ determinants. It is very intriguing that methods which seem to be very powerful in arithmetic geometry appear to be also intimately connected with the mathematical structure of string theories. Whether these mathematical structures reflect some fundamental property of string theory and not just of its perturbation expansion, is something that remains to be clarified.

In the following we will describe what bosonization says about the partition function and correlation functions of generalized ghost systems.

Following the treatment of Ref. 38, we consider a first-order ghost system whose dynamics is described by the action

$$S = \int_\Sigma b\bar{\partial}c + c.c. \tag{25}$$

where b is a λ-differential and c is a 1-λ-differential. Note that $b\bar{\partial}c$ is a (1,1)-form and therefore can be integrated over Σ without any additional factor. Geometrically the field c is a section of $L^{2-2\lambda}$ where L is a twisted spin bundle. In general the $\bar{\partial}$ operator may have a kernel and/or a co-kernel. In order to obtain a non-vanishing partition function we have to insert the appropriate number of b and c fields to soak up the zero modes. This also follows form the fact that the ghost number current bc has an anomaly leading to a net violation of ghost charge $\Delta Q = k = (2\lambda-1)(g-1)$, as required by the Riemann-Roch theorem (for more details, see for example 38). In particular, for $\lambda = \frac{1}{2}$, no insertions are required; for $\lambda = 1$, we need one insertion of c and g insertions of b; and k insertions of b for $\lambda > 1$. Thus, if ψ_i, $i = 1,\dots,k$ denotes a basis of holomorphic λ-differentials (λ>1), and ω_i, $i = 1,\dots,g$ is a basis of holomorphic one-forms (Abelian differentials), we can write the partition functions to be computed as

$$Z(\lambda=1/2) = \det \bar{\partial}_L^+ \bar{\partial}_L$$

$$Z(\lambda=1) = \int db\,d\bar{b}\,dc\,d\bar{c} \prod_{i=1}^{g} b\bar{b}(P_i)\, c\bar{c}(Q)\, e^{-S} =$$

$$= \frac{\det \omega_i(P_j) \wedge \det \bar{\omega}_i(P_j)}{\det \langle \omega_i | \omega_j \rangle} \frac{\det' -\nabla^2}{\int \sqrt{g}}$$

$$Z(\lambda>1) = \int db\,d\bar{b}\,dc\,d\bar{c} \prod_{i=1}^{k} b\bar{b}(P_i)\, e^{-S} =$$

$$= \frac{\det \psi_i(P_j) \wedge \det \bar{\psi}_i(P_j)}{\det \langle \psi_i | \psi_j \rangle} \det' \bar{\partial}^+_{L^{2-2\lambda}} \bar{\partial}_{L^{2-2\lambda}} \tag{26}$$

Since we have inserted b fields, Z(λ) should be considered as a differential form on Σ^k. In (26), the factor of $\sqrt{g}$ represents the volume of the surface for the metric chosen. The prime means that zero modes have been removed. The computation of the determinants (26) is greatly simplifed from a mathematical point of view if we choose the Arakelov metric on Σ. Since the integrand in the string path integral is conformally invariant, it does not matter which metric we choose. There are two metrics which have proved useful in string theory: (i) the Poincaré metric, obtained from the general uniformization theorem[42] stating that any Riemann surface of genus $g > 1$ can be represented as the upper half plane divided by a hyperbolic fuschian group; a discrete subgroup of SL(2,R), the isometry group of the upper half plane with the Poincaré metric $ds^2 = dzd\bar{z}/(\mathrm{Im} z)^2$. With this representation of the Riemann surface, one can use the Selberg trace formula to obtain useful expressions for the determinants in (26)[43]. (ii) The Arakelov metric: on any Riemann surface of genus $g > 0$, there are g holomorphic one-forms, ω_i $\omega_i(z)dz$, $ik = 1,\dots,g$. A canonical basis for the homology of the surface is given by a collection of one-cycles a_i, b_i, $i = 1,\dots,g$ whose intersection pairing is given by

$$(a_i, a_j) = (b_i, b_j) = 0$$

$$(a_i, b_j) = -(b_i, a_j) = \delta_{ij} \tag{27}$$

a convenient normalization of the basis of Abelian differentials is to require that

$$\int_{a_i} \omega_j = \delta_{ij} \tag{28}$$

then the periods over the b-cycles are determined

$$\int_{b_i} \omega_j = \Omega_{ij} \tag{29}$$

Ω is known as the period matrix of the Riemann surface for the given marking (choice of a canonical homology basis), and it is a symmetric matrix with positive definite imaginary part. Using the Abelian differentials we can construct the (1,1)-form

$$\mu = \frac{i}{2g} \sum_{ij} \omega_i \, (\mathrm{Im}\,\Omega)^{-1}_{ij} \, \bar{\omega}_j \tag{30}$$

satisfying

$$\int_{\Sigma} \mu = 1 \tag{31}$$

Given a holomorphic line bundle ξ over Σ, a metric on it is admissible if its curvature is proportional to μ. Thus if s is a section of ξ, and $\|s\|^2$ its norm, the curvature R_ξ is given by

$$R_\xi = \bar{\partial}\partial \log \|s\|^2 \propto \mu. \tag{32}$$

Using the Arakelov metric on Σ, we can construct the Arakelov Green's function G(P,Q) satisfying

$$\begin{gathered} \partial\bar{\partial} \log G(P,Q) = i\pi\mu(P) - i\pi\delta(P,Q) \\ \int_{\Sigma} \log G(P,Q)\, \mu(P) = 0 \qquad G(P,Q) = G(Q,P) \end{gathered} \tag{33}$$

the explicit form of the determinants (26) is given in terms of G(P,Q).

Bosonization is the statement that the first-order field theories (25) and (26) can be replaced by equivalent bosonic field theories. In Ref. 30, it was shown that the local properties of the weight λ-system can be reproduced by the scalar action

$$S[\varphi] = 4\pi i \int \partial\varphi\bar{\partial}\varphi + (2\lambda - 1) \int R_{K^{-1}} \varphi \tag{34}$$

the first term is the standard kinetic term, and the second term accounts for the anomaly in the ghost number current. $R_{K^{-1}}$ is the curvature of the holomorphic tangent bundle. K is the canonical line bundle; the bundle of holomorphic one-forms. The curvature in (34) is normalized so that $\int R = 4\pi i(g-1)$. The bosonization prescription for ghost bilinears is then

$$\begin{aligned} b\bar{b} &= \,:\exp 4\pi i\varphi: \\ c\bar{c} &= \,:\exp -4\pi i\varphi: \end{aligned} \tag{34}$$

and the normal ordering prescription cancels the co-ordinate dependence so that b and c are well-defined forms. Normal ordering requires one to define the Green's functions at coincident points. In the Arakelov metric, the prescription is[39]:

$$:\log G(P,Q): = \lim_{P\to Q} \{ G(P,Q) - 2\lambda \log |z(P) - z(Q)| - (1-2\lambda) \log d(P,Q) \} \tag{35}$$

z is a local co-ordinate in a neighbourhood of P and d(P,Q) is computed with the Arakelov metric. On a multiply connected surface, the field ϕ is not single-valued and it can have non-trivial shifts around the homology cycles. Since mass terms and currents should be well defined on the surface, (34) forces the shifts of ϕ to be either integral or half-integral. In the evaluation of the functional integral for the scalar field (34), we can split the field ϕ into a single-valued piece ϕ and an instanton $\phi(n,m)$. The instanton is constructed such that $d\phi(n,m)$ is a globally defined harmonic one-form with n_i periods over the a-cycles and m_i periods over the b-cycles, n_i, m_i integers or half-integers, $i = 1.,,,,.g$. To evaluate the second term in (34) in an instanton section we must define the multiple-valued field ϕ by choosing a system of curves intersecting in a single point A and being homologous to the homology basis chosen. Such a choice lets us cut open the surface to obtain a 4g-sided polygon Σ_c. Choosing now a base point P_0, the field

$$\varphi(n,m) = \int_{P_0}^{P} d\varphi(n,m) \tag{36}$$

is well defined on Σ_c, and we may evaluate the action (34). In doing so, one finds that the bosonic action and path integral depend explicitly on the marking. This is unacceptable because for non-chiral ghost systems [as in (25)], the partition function does not depend on the choice of homology basis (it is modular invariant): Therefore we must change the action (34) in order to achieve marking independence. In fact the new terms to be added to (34) can be uniquely determined by very plausible requirements. Namely[39]:

(i) independence of A and P_0,

(ii) independence of the representative cycles on the homology classes of our homology basis,

(iii) the conformal anomaly and local conformal properties should not be changed,

(iv) modular invariance of the action for multivalued fields (independence of the homology basis),

(v) factorization of the instanton partition function.

Since the instanton quantum numbers are only related to the homology basis, one would expect that when the surface Σ is split into two pieces Σ_1, Σ_2 of genus g_1, g_2 (resp.), $g_2, g_2 > 0$ by shrinking to zero a homologically trivial cycle, the instanton partition function would also factorize. Before writing down the modified action obtained after use has been made of (i)-(v), we need to introduce a bit more notation. After a marking has been chosen there is a preferred spin structure L_Δ associated to the vector of Riemann constants[44], and it is natural to use this spin bundle to parametrize the line bundles whose sections are represented by b and c. Since two holomorphic line bundles with the same first Chern number only differ by a flat line bundle[45], we can write $(L \otimes L_\Delta^{-1})^{2-2\lambda} = F(\theta_1, \theta_2)$, where F is a holomorphic line bundle with holonomy $2\pi i\theta_1$, $-2\pi i\theta_2$ around the a and b-cycles, respectively. After all these preliminaries we can finally write down the complete scalar action

$$S[\varphi] = 4\pi i \int \partial\varphi \bar{\partial}\varphi + (2\lambda - 1) \int R_\kappa^{-1} \varphi$$
$$+ \sum_{\kappa=1}^{g} \left\{ \int_{a_\kappa} d\varphi \, f(b_\kappa) - \int_{b_\kappa} d\varphi \, f(a_\kappa) \right\}$$
$$+$$

$$+\, 4\pi i \sum_{K=1}^{g} \int_{a_K} d\phi \int_{b_K} d\phi \tag{37}$$

where $f(\gamma)$ is a function defined for any closed curve and it is given by twice the holonomy of the admissible connection for the line bundle $L^{+2-2\lambda} \otimes L_\Delta^{-1}$

$$f(\gamma) = 2 h(\gamma; L^{2-2\lambda} \otimes L_\Delta^{-1}) \tag{38}$$

By using (37) and (38) and the prescription (34) to evaluate (26) (assuming that bosonization works) we obtain the following formulae for the determinants appearing in (26)[39], computed with the Arakelov metric

$$\det \bar\partial_L^+ \bar\partial_L = \left(\frac{\det' -\nabla^2}{\int\sqrt{g} \, \det \operatorname{Im}\Omega} \right)^{-1/2} \| \vartheta(\theta_1 + \Omega\theta_2 | \Omega) \|^2 \tag{39a}$$

$$\left(\frac{\det' -\nabla^2}{\int\sqrt{g} \, \det \operatorname{Im}\Omega} \right)^{3/2} = \frac{\prod_{i<j} G^2(P_i, P_j) \, \| \vartheta(z_{\lambda=1} | \Omega) \|^2}{\| \det \omega_i(P_j) \|^2 \, \prod G(P_i, Q)^2} \tag{39b}$$

$$\frac{\det \bar\partial_{L^{2-2\lambda}}^+ \bar\partial_{L^{2-2\lambda}}}{\det \langle \psi_i | \psi_j \rangle} = \left[\frac{\det' -\nabla^2}{\int\sqrt{g} \, \det \operatorname{Im}\Omega} \right]^{-1/2} \frac{\prod_{i<j} G(P_i, P_j)^2 \, \| \vartheta(z_\lambda | \Omega) \|^2}{\| \det \psi_i(P_j) \|^2} \tag{39c}$$

where Θ is the Riemann Θ-function for the Riemann surface given by[44]:

$$\vartheta(z|\Omega) = \sum_{(n_i) \in \mathbb{Z}^g} \exp(i\pi n_i \Omega_{ij} n_j + 2\pi i n_i z_i)$$

and

$$\| \vartheta(z|\Omega) \|^2 = e^{-2\pi (\operatorname{Im} z)_i (\operatorname{Im}\Omega)^{-1}_{ij} \operatorname{Im} z_j} \, | \vartheta(z|\Omega) |^2$$

$$z_{\lambda=1} = I\left(\sum_{i=1}^{g} P_i - Q \right) - \Delta$$

$$z_\lambda = I\left(\sum_{i=1}^{K} P_i - (2\lambda - 1)\Delta \right) + \theta_1 + \Omega\theta_2$$

$$I\left(\sum_{i=1}^{n} P_i \right) = \sum_{i=1}^{n} \int_{P_0}^{P_i} \omega$$

and in the very last formula we have assembled the Abelian differentials to form a vector. Similar formulae can be obtained for correlation functions of an arbitrary number of insertions of b and c fields[38]. As pointed out at the beginning of this section, the rigorous proof of these bosonization formulae[39] makes use of Quillen's theorem[40], some of Faltings' constructions[41], and the very particular property of moduli space that any globally defined holomorphic function is a constant[35]. The results expressed in Eq. (39) can be used to write fairly explicit formulae for the string partition and correlation functions in high orders of perturbation theory, and thus analyze issues concerning finiteness as well as general properties of string perturbation theory with the help of rather powerful methods in algebraic geometry. Perhaps the most outstanding recent result of the interplay between algebraic geometry and string theory is Manin's[46] evaluation of the string integrand for the bosonic string in a surface of arbitrary genus. It should also be said, however, that if one is interested in obtaining definite numbers for a given string amplitude one is still left with the rather difficult problem of performing the integration of expressions similar to those in (36) over moduli space. Nevertheless, many qualitative properties of the theory can be formulated in terms of the partition and correlation functions at generic points in moduli space. Even more interesting is the possibility suggested by Friedan and Shenker[47] in their approach to quantum string theory that these beautiful structures which appear in perturbation theory, do capture some of the fundamental properties of string theory. These aspects of the theory and their connection with string field theory are being investigated vigorously, and hopefully we will have a more clear picture not too far in the future.

ACKNOWLEDGEMENTS

I would like to thank Prof. A Zichichi for his invitation to participate in this School and for his kind hospitality.

REFERENCES

1. For a review and references to the original literature, see: L. Alvarez-Gaumé - "An Introduction to Anomalies" HUTP 85/A052, to appear in the Proceedings of the Erice School in Mathematical Physics; G. Velo and A. Wightman, eds. (Plenum Press, 1986).

2. E. Witten - Phys.Lett. 117B (1982) 324; Comm.Math.Phys. 100 (1985) 197.

3. E. Witten - in Proceedings of the Argonne Symposium on Anomalies Geometry and Topology. W.A. Bardeen and A. White, eds. (World Scientific, 1985).

4. S.L: Adler - Phys.Rev. 177 (1969) 2426, and in "Lectures in Elementary Particles and Quantum Field Theory", S. Deser et al., eds. (MIT Press, 1970);
J.S. Bell and R. Jackiw - Nuovo CImento 60A (1969) 47.
See also: R. Jackiw - in "Lectures on Current Algebra and its Application", (Princeton U. Press, 1972).

5. G. 't Hooft - Phys.Rev.Lett. 37 (1976) 8;
C. Callan, R. Dashen and D. Gross - Phys.Lett. 63B (1976) 334;
R. Jackiw and C. Rebbi - Phys.Rev.Lett. 37 (1976) 172.

6. R. Delbourgo and A. Salam - Phys.Lett. 40B (1972) 381;
T. Eguchi and P. Freund - Phys.Rev.Lett. 37 (1976) 1251.

7. D.J. Gross and R. Jackiw - Phys.Rev. D6 (1972) 477;
C. Bouchiat, J. Iliopoulos and Ph. Meyer - Phys.Lett. 38B (1972) 519;
H. Georgi and S.L: Glashow - Phys.Rev. D6 (1972) 429.

8. P. Langacker - Physics Reports 72 (1981) 185.

9. P.H. Frampton and T.W. Kephart - Phys.Rev.Lett. 50 (1983) 1343, 1347;
P.K. Townsend and G. Sierra - Nucl.Phys. B222 (1983) 493;
B. Zumino, W.Y. Shi and A. Zee - Nucl.Phys. B239 (1984) 477.

10. L. Alvarez-Gaumé and E. Witten - Nucl.Phys. B234 (1983) 269.

11. M. Green and J. Schwarz - Phys.Lett. 149B (1984) 117;
M. Green, J. Schwarz and P. West - Nucl.Phys. B254 (1984) 377.

12. D.J: Gross, J. Harvey, E. Martinec and R. Rohm - Phys.Rev.Lett. 54 (1985) 502; Nucl.Phys. B256 (1985) 253; B267 (1986).

13. M.F. Atiyah and I.M. Singer - Ann. of Math. 87 (1968) 485, 546; 93 (1971) 1, 119, 139;
M.F. Atiyah and G.B. Segal - Ann. of Math. 87 (1968) 531.

14. R. Jackiw, C. Nohl and C. rebbi - in "Particles and Fields", D. Boch and A. Kamal, eds. (Plenum Press, NY, 1978);
N.K. Nielsen, H. Roemer and B. Schroer - Nucl.Phys. B136 (1978) 478.

15. L. Alvarez-Gaumé and P. Ginsparg - Nucl.Phys. B243 (1984) 449.

16. M.F. Atiyah and I.M. Singer - Proc.Nat.Acad.Sci. U.S.A. 81 (1984) 2597.

17. S.S. Chern - "Complex Manifolds without Potential Theory" (Van Nostrand, 1967).

18. See Ref. 1 for a thorough discussion.

19. For very clear discussions, see:
B. Zumino - in "Relativity, Groups and Topology - II", Proceedings of the Les Houches Summer School, B.S. DeWitt and R. Stora, eds. (North Holland, 1984);
R. Stora - in "PRogress in Gauge Theories", Proceedings of the 1984 Cargèse School, G. 't Hooft et al., eds. (Plenum Press, 1984).

20. L.D. Faddeev - Phys.Lett. 145B (1984) 81;
L.D. Faddeev and S. Shatashvili - Teor. and Math.Phys. 60 (1984) 206;
B. Zumino - Nucl.Phys. B253 (1985) 477;
R. Jackiw - MIT Preprint CTP 1298 (198), to appear in Comm.Nucl. Part.Phys.;
P. nelson and L. Alvarez-gaumé - Comm.Math.Phys. 99 (1985) 103.

21. G. 't Hooft - in "Recent Progress in gauge Theories", G. 't Hooft et al., eds. (Plenum Press, N.Y., 1980).

22. W.A. bardeen and B. Zumino - Nucl.Phys. b244 (1984) 241.

23. A.N. Redlich - Phys.Rev.Lett. 52 (1984) 1.
See also Ref. 10.

24. L. Alvarez-Gaumé, S. Della Pietra and G. Moore - Ann.Phys. 163 (1985) 288.

25. A. Niemi and G. Semenoff - Physics Reports 136 (1986) 100.

26. M.F. Atiyah, V.I. Patodi and J.M. Singer - Proc.Camb.Phil.Soc. 77 (1975) 43; 78 (1975) 405; 79 (1976) 71.

27. J. Schwarz - Physics Reports 89 (1982) 223; and Proceedings of the Santa Barbara Workshop on Superstrings, M. Green and D. Gross, eds. (World Scientific, 1986).

28. F. Gliozzi, J. Scherk and D. Olive - Nucl.Phys. B122 (1977) 253.

29. See Ref. 3.

30. J. Birman - "Braids, Links and Mapping Class Groups", Lecture Notes in Mathematics (Princeton University Press, 1974).

31. R. Stong - "Lectures on Cobordism Theory", Lecture Notes in Mathematics (Princeton University Press, 1968).

32. L. Alvarez-Gaumé, P. Ginsparg, G. Moore and C. Vafa - HUTP 86/A013; L. Dixon and J. Harvey - Princeton Preprint

33. B.A. Schellekens and N. Warner - CERN Preprints TH. 4464 and 4465 (1986).

34. A.M. Polyakov - Phys.Lett. 103B (1981) 207, 211;
D. Friedan - in "Recent Advances in Field Theory and Statistical Mechanics", J.B. Zuber and R. Stora, eds. (Elsevier, 1984);
O. Alvarez - Nucl.Phys. B216 (1983) 125;
J. Polchinski - Comm.Math.Phys. 104 (1986) 37;
G. Moore and P. Nelson - Nucl.Phys. B266 (1986) 58;
E. D'Hoker and D. Phong - Nucl.Phys. B269 (1986) 206.

35. H. Harris - in "Proceedings of the International Congress of Mathematicians", Warszawa, Poland, C. Olech and Z. Ciesielski, eds. (Elsevier, Amsterdam, 1984).

36. A. Belavin and V.I. Knizhnik - "Algebraic Geometry and the Geometry of Quantum Strings", Landau Institute Preprint (to appear in Nucl.Phys.);
R. Catenacci, M. Cornalba, M. Martellini and C. Reina - "Algebraic Geometry and Path Integrals for Closed Strings", Pavia Preprint;
J.B. Bost and T. Jolicoeur - Saclay PhT/86-28 (1986).

37. L. Alvarez-Gaumé, G. Moore and C. Vafa - Comm.Math.Phys. 106 (1986) 40;
See also: J.B. Bost and P. Nelson - Phys.Rev.Lett. 57 (1986) 795.

38. D. Friedan, E. Martinec and S. Shenker - Nucl.Phys. B271 (1986) 93.

39. L. Alvarez-Gaumé, J.B. Bost, G. Moore, P. Nelson and C. Vafa - HUTP 86/A036 (1986) (to appear in Phys.Lett. B) and in preparation.

40. D. Quillen - Funct.Anal.Appl. 19 (1986) 31.

41. G. Faltings - Ann. of Math. 119 (1984) 387.

42. H. Farkas and J. Kra - "Riemann Surfaces", Springer (1980).

43. E. D'Hoker and D. Phong - Comm.Math.Phys. 104 (1986) 537.

44. D. Mumford - "Tata Lectures on Theta", Birkhauser (1983).

45. G. Gunning - "Introduction to Riemann Surfaces", Princeton Lecture Notes in Mathematics (Princeton University Press, 1966).

46. Yu. Manin - JETP Lett. 43 (1986) 204.

47. D. Friedan and S. Shenker - Chicago Preprints EFI-86-18A,B (1986).

DISCUSSIONS I

CHAIRMAN: L. Alvarez–Gaumè

Scientific Secretaries: H. Livne, K. Meissner, Y. Shamir and A. Shapere

– *A. Pasquinucci:*

Why are local anomalies in 2n dimensions described by (2n +2)–forms?

– *L. Alvarez–Gaumè:*

Consider the Dirac operator D_A as a function of the gauge field $\mathcal{A}$. Let $\mathcal{A}$ be the space of all gauge fields and $\mathcal{G}$ the space of all gauge transformations. Now $\mathcal{A}$ is topologically trivial if I take two gauge configurations A_1 and A_2, the I can always draw a line in $\mathcal{A}$, $A_t = (1-t)A_1 + tA_2$, which connects A_1 and A_2 as t goes from 0 to 1. Hence $\mathcal{A}$ is contractible. But since $\mathcal{G}$ has a large amount of topology, the space of gauge–configurations $\mathcal{A}/\mathcal{G}$ is also topologically nontrivial. In fact

$$\pi_1(\mathcal{A}/\mathcal{G}) = \pi_0(\mathcal{G})$$

$$\pi_2(\mathcal{A}/\mathcal{G}) = \pi_1(\Im)$$

and so on. To understand the last equation (and its relationship to local anomalies), let's try to construct a nontrivial 2–sphere in $\pi_2(\mathcal{A}/\mathcal{G})$ from an element of $\pi_1(\mathcal{G})$. Consider a gauge field A and a one–parameter family $g(\theta)$ of gauge transformations, so that $g(\theta) : S^1 \times S^{2n} \to G$. Now construct a disc D in $\mathcal{A}$ whose elements are $A(t,\theta) = tA^{g(\theta)}, 0 \leq t \leq 1$. Then when projected down to $\mathcal{A}/\mathcal{G}$, the boundary of D goes to a point and D becomes a 2–sphere. This 2–sphere is noncontractible if $g(\theta)$ is topologically nontrivial in $\pi_1(\mathcal{G}) \sim \pi_{2n+1}(G)$. For $G = SU(N)$ in 4 dimensions, $\pi_5(SU(N)) = Z$ and so there are many nontrivial ways to make this construction.

The relationship of $\pi_2(\mathcal{A}/\mathcal{G})$ to the local anomaly is as follows. Over $\mathcal{A}/\mathcal{G}$, we compute the effective action. This gives us a complex number for each A, and we obtain a complex line bundle over $\mathcal{A}/\mathcal{G}$. Nontriviality of this line bundle means that $\Gamma[A]$ cannot be globally defined as a single–valued function on $\mathcal{A}/\mathcal{G}$, and hence that the theory is anomalous. We can find all the topological information contained in the line bundle by looking at its restriction to noncontractible 2–spheres in $\mathcal{A}/\mathcal{G}$.

Now we can see why we compute the anomaly in 2n + 2 dimensions. $A(t,\theta,x)$ is effectively a gauge field over the 2n + 2 dimensional space $S^2 \times S^{2n}$, with field strength F, and the topological number of the element of $\pi_2(\mathcal{A}/\mathcal{G})$ that it corresponds to is proportional to

$$\int_{S^2\times S^{2n}} T_r F^{n+1}$$

– *A. Pasquinucci:*

In four dimensions, the non–abelian anomaly is described by a six–form. How does one recover the standard expression for the anomaly from this six–form.

– *L. Alvarez-Gaumè:*

Global anomalies like the U(1) anomaly in QCD are always given by the standard index theorem in 2n dimensions. But for local anomalies of four–dimensional gauge theories, you must start with

$$Tr\ F^3 = 3d\int_0^1 dt\ Tr\ A(tdA + t^2A^2)^2 \equiv dQ_5^0$$

where Q_5^0 is the Chern–Simons 5–form. Now make a gauge transformation $A \rightarrow A + Dv$ where $Dv \equiv dv + [A, v]$. Then

$$\delta Q_5^0 = dQ_4^1$$

where

$$Q_4^1 \equiv Str\ vd(AdA + \frac{1}{2}A^3)$$

This is explicitly related to the resulting change in the effective action by

$$\delta\Gamma[A] = \frac{1}{24\pi^2}\int_{S_4} Q_4^1$$

– *Z. Bern:*

Can you give a deep physical interpretation of the Atiyah–Singer index theorem and why it gives the anomaly?

– *L. Alvarez-Gaumè:*

The index theorem has both an Euclidean and a Hamiltonian formulation. Let us look at the Dirac operator and try to build up a gauge invariant theory satisfying Gauss' law and then we can see how to interpret the result in terms of the Euclidean path integral formulation. To understand the anomalies, we only have to look at the gauge field as an external background, we do not have to consider it as a dynamical object. So we start with the Hamiltonian of a fermion in the presence of a background gauge field (take $A_0 = 0$). Then

$$H = \int d^3x \lambda . \not{D}(A) \lambda$$

For each gauge field, we have a Hilbert space $\mathcal{H}_{\mathcal{A}}$ on top of it. As usual, we can expand λ in terms of normal modes

$$\lambda(x) = \sum_{E>0} a_E \psi_E(x) + \sum_{E<0} b_E \Psi_E(x)$$

We now want to go from the Hilbert space $\mathcal{H}_{\mathcal{A}}$ to the Fock space $\mathcal{F}_{\mathcal{A}}$. The Fock vacuum is $|0_A>$ obtained as usual by filling the Dirac sea. Then we get the rest of $\mathcal{F}_{\mathcal{A}}$ by acting with the appropriate creation or annihilation operators. Since we eventually want to quantize A, we need a second quantized theory that is consistent with Gauss' law. This means that the Fock vacuum $|0_{Ag}>$ for a gauge transformation g of A must be some unitary transformation U_{A_g} which can be constructed in terms of the fermionic currents of the Fock vacuum $|0_A>$. Obviously, the $U's$ should satisfy a composition law

$$U_{A_1 g_1} U_{A_1 g_2} = U_{A_1 g_1 g_2}$$

That is, we do not want any projective phases to appear. This is a Hamiltonian statement of Gauss' law, which must be satisfied if we want to quantize A. To see if it is satisfied, we can do an adiabatic calculation. The space of Fock vacua over the space of inequivalent 3–dimensional gauge configuration $\mathcal{A}^{(3)}/\mathcal{G}^{(3)}$ is a line bundle. We want to check if the line bundle twists in a non trivial way. We do this by looking at two–spheres in $\mathcal{G}^{(3)}$ corresponding to three–spheres in $\mathcal{A}/\mathcal{G}^{(3)}$. That is, if $g(\theta_1 \phi) : S^2 \to \mathcal{G}$, then consider a three–ball B in A whose boundary is the set of $A^{g(\theta,\phi)}$. These projects to a three–sphere in $\mathcal{A}^{(3)}/\mathcal{G}^{(3)}$. Now

$$\pi_3(\mathcal{A}/\mathcal{G})^{(3)} = \pi_2(\mathcal{G}^{(3)}) = \pi_5(G)$$

and so if $\pi_5(G)$ is nontrivial, we should look at the corresponding nontrivial 3–spheres S in $\pi_3(\mathcal{A}^{(3)}/\mathcal{G}^{(3)})$. If S is non–contractible, then you can show using the index theorem that there must be points A_1 and A_2 on the boundary of B such that the line

$$A_t = tA_1 + (1-t)A_2$$

joining them contains a point where two of the levels of the Weyl operator $D(A_t)$ cross at zero. So, empty Fock space states ($E > 0$) become filled. If you do this more carefully, you can show that the net flux of gauge charge is non zero and thus that the theory is anomalous.

– *P. Fisher:*

How do you show that the standard model is anomaly free?

– *L. Alvarez-Gaumè:*

We want to show that the symmetrized trace $strT_1^aT_2^bT_3^c$ vanishes for all possible choices of external gauge currents in the triangle diagram. If T_i^a is a generator of the group G_i, then we must check this when $G_1G_2G_3$ equals

$$
\begin{aligned}
(i)&\quad SU(3)^3\\
(ii)&\quad SU(2)^3\\
(iii)&\quad U(1)^3\\
(iv)&\quad U(1)\quad SU(2)^2\\
(v)&\quad U(1)\quad SU(3)^2
\end{aligned}
$$

All cases with just one factor of $SU(2)$ or $SU(3)$ give no anomaly, since the trace of generator of a simple group is zero.

For case (i) the quantum numbers of $SU(2)\times U(1)$ are irrelevant. The theory is just a theory of Dirac fermions in the 3 of $SU(3)$, so it is anomaly–free. $SU(2)^3$ is also fine because $SU(2)$ has only real and pseudoreal representations. A representation is real or pseudoreal if $X^t = -S\ S^{-1}$ for X in the representation of the Lie Algebra. S is symmetric for a real representation and antisymmetric for a pseudoreal one. In either case, $TrX^3 = 0$.

Case (iii) has already been checked. The remaining two cases can easily be checked by explicit computation.

– *A. Shapere:*

How do you see that there are no global anomalies in the standard model?

– *L. Alvarez-Gaumè:*

In four dimensions, global anomalies come from

$$\pi_0(\mathcal{G}) = \pi_4(G)$$

Now $\pi_4(G) = 0$ except for $Sp(2n)$. In particular, $\pi_4(Sp(2)) = \pi_4(SU(2)) = Z_2$. Witten showed that for $SU(2)$ gauge theories with an odd number of Weyl fermions in the fundamental representation, the effective action changes sign when you make a nontrivial gauge transformation corresponding to the generator of the nontrivial element of $\pi_4(SU(2))$. We are lucky in the standard model because we have an even number of doublets with respect to $SU(2)$. If you did not have an even number of doublets, you would suspect that either there was another fermion doublet or else a rather subtle Wess–Zumino term involving scalar fields which would serve to cancel the phase in the fermion determinant. In six dimensions, you can find amusing things like $\pi_6(SU(3)) = Z_6$, leading to more complicated phases. Even more exotic possibilities may occur in higher dimensions but they have not been thoroughly studied.

– *S. Carlip:*

What is so bad about a theory with global anomalies? Can't you simply say that the actual invariance group is the connected component of the gauge group?

– *L. Alvarez-Gaumè:*

An analogous and closely related question would be why you cannot define a path integral for gauge fields with only zero topological charge. The result would be bad clustering properties: a field with total topological charge zero can have a charge of + 100 here and - 100 at the other side of the room, so if you want your theory to be local you need to include configurations with non zero charges. Similarly you can't throw out disconnected gauge transormations – a gauge transformation homotopic to the identity can look (almost) like two separated topologically twisted gauge transformations.

– *P. Vecsernyes:*

What happens with anomalies in odd dimensions?

– *L. Alvarez-Gaumè:*

In odd dimensions, an anomaly appears because you cannot regularize the effective action in a way which preserves both gauge invariance and, say CP or PT invariance. That is, the imaginary part of the effective action for an odd–dimensional gauge theory is $e^{i\frac{\pi}{2}\eta(0)}$, and the eta–invariant $\eta(0)$ violates PT inversion. This is because PT and CP exchange the positive and negative eigenvalues of the Dirac operator, and hence change the sign of $\eta(0)$. Moreover, if you were to work on an arbitrary manifold admitting a time–reversal isometry, then it is inevitable that there will be such an anomaly. But in principle, this is no problem, because you are only violating a discrete symmetry. This anomaly is responsible in some cases for the phenomenon of fermion fractionalization in odd dimensions.

– *S. Carlip:*

The gauge groups $E_8 \times E_8$ and Spin $(32)/Z_2$ are picked out by two very different arguments – ten dimensional gauge anomaly cancellation, and modular invariance of the heterotic string. What is the connection between modular invariance of a string theory and gauge invariance of the corresponding low–energy field theory?

– *L. Alvarez–Gaumè:*

Does modular invariance guarantee anomaly cancellation? It is not yet completely clear. It should in principle be possible to check this by looking at parametrized families of string effective actions with background gauge and gravitational fields. This is what we do in an ordinary gauge theory when compute anomalies using the index theorem. One technical difficulty with this approach involves the sum over spin structures – it is not explicitly known, in general, how to perform this sum in a way consistent with modular invariance.

A connection between modular invariance and anomaly cancellation should not be surprising. In the σ–model approach to string theory, the partition function is also the generating function for the string S–matrix. Thus, if you think about the first–quantized formulation from that point of view, it is reasonable to expect that modular invariance has something to do with anomaly cancellation, because modular invariance is the only thing that guarantees that the spectrum you find is going to be consistent with the quantization rules.

– *Y. Shamir:*

I want to ask about the use of index theorem for various calculations. Since the index theorem is proved for compact spaces is it automatic that the result applies to unbounded spaces or does one need extra necessary conditions?

– *L. Alvarez–Gaumè:*

If you look at the local anomaly, namely the anomaly you get from Feynman diagrams, in a sense it doesn't really matter if you use the index theorem for a compact manifold or for a non compact manifold. Because, for a local anomaly you are going to reproduce a local function, and the index theorem is really a trick to get the same answer that you would obtain by using Feynman diagrams. The index theorem gives you the correct answer regardless of what the gauge group or the topology of the space is. For example, these polynomials that you derive from the index theorem also give you the right answer for a $U(1)$ gauge theory with Weyl fermion and of course $\pi_5(U(1)) = 0$. In a sense, what you do is to use the possibility of having a nontrivial topology for some particular gauge groups and then you use the index theorem to derive what the anomaly should be. If you want to look at global anomalies for fermion fractionalization or things of this type which also involve the index theorem then, of course, having boundaries and which boundary conditions are imposed play a role. But this has to be done more on a case by case basis.

DISCUSSION II

– *U. Loew:*

How is the determinant of an operator defined?

– *L. Alvarez-Gaumè:*

Let us first consider an operator $A : V \to V$ whose eigenvalues are all well-defined. If the spectrum is defined by

$$A\Psi_n = \lambda_n \Psi_n$$

Then we want *det A* to be the product of the eigenvalues λ_n. We regularize this product using ζ–functions that is, construct first

$$\zeta_A(s) = \sum_n \frac{1}{\lambda_n^s}$$

and then define

$$det\ A \equiv \bar{e}^{\zeta'_A(0)}$$

What happens for operators that do not have an interpretation in terms of eigenvalues? First consider the finite dimensional case. If A maps V into itself, then a fancy way to define *det A* which generalizes easily is as the volume into which A maps a unit volume of V. In more mathematical terms, *det A* is the mapping from the highest exterior power of V into itself induced by $A : V \to V$. Thus

$$det\ A : \Lambda^{max} V \to \Lambda^{max} V$$

takes $v_1 \wedge \cdots \wedge v_n \to Av_1 \wedge \cdots \wedge Av_n = (det\ A) v_1 \wedge \cdots \wedge v_n$, *det A* is a 1–dimensional linear mapping in $\Lambda^{max} V \otimes (\Lambda^{max} V)^*$. Things are not so simple when A maps V into a different space W. In this case, we define *det A* to be a one–dimensional vector space.

$$det\ A = (\Lambda^{max} W) \otimes (\Lambda^{max} V)^*$$

For a general infinite dimensional operator, there is no such invariant notion of a determinant. But for operators with a finite dimensional kernel and cokernel, known as Fredholm operators, I can define a determinant in a way analogous to the case of finite dimensional operators. We use the exact sequence. $0 \to ker\ A \to V \xrightarrow{A} W \to cok\ A \to 0$. This is an exact sequence because the image of any one of the arrows is equal to the kernel of the next arrow. Thus, the composition of any two arrows gives zero. One property of exact sequences is that the alternating ratio of volumes is equal to 1. If V and W are finite–dimensional, this means that

$$\Lambda^{max} W \otimes (\Lambda^{max} V)^* = \Lambda^{max} cok\ A \otimes (\Lambda^{max} ker\ A)^*$$

This gives us a natural definition for the determinant of a Fredholm operator which reduces to the definition we have already given in the finite dimensional case, namely,

$$det\ A \equiv \Lambda^{max} cok\ A \otimes (\Lambda^{max} ker\ A)^*$$

but this is not enough, because we want to get a number out of this one–dimensional vector space. We need to define a norm for the vector space, which will in general depend on the particular physical problem in question. In certain cases, this norm is what one calls the determinant of the operator. For the case of the determinant of the Weyl operator * on a Riemann surface, everything is holomorphic and the complex structure gives a natural metric on the space *det*. In particular, when we consider the operators appearing in the first quantized formulation of string theories, there is a natural way of associating a holomorphic function to the combined product of determinants.

– *A. Shapere:*

What is involved in checking finitess of string theory amplitudes? In particular how does one check that there are no divergences which come from worldsheets with nodes?

– *L. Alvarez–Gaumè:*

The complex structure of moduli space allows us to use powerful methods from algebraic geometry in addressing the question of finitess. We have discussed the holomorphic dependence of the determinant.

$$det\bar{\partial}_q(t)$$

on the coordinates of moduli space M. On the boundary of M, the corresponding Rieman surfaces have nodes, which may lead to poles in the determinant as a function of t. As we discussed, the proper way to view the determinat is as a holomorphic line bundle over M. And all the interesting information about a line bundle comes from its divisor, which is essentially the set of points where the determinant has a pole or a zero. It is hard to calculate the divisor of our determinant line bundle directly. But we can use the fact that this is a holomorphic line bundle over a complex space to obtain the divisor indirectly by using the Grothandieck– Riemann–Roch theorem from algebraic geometry. A non–trivial divisor will come from the nodes where the Riemann surfaces degenerate. We must check that the various poles cancel. For example, the wrong sum over spin structures may lead to second order poles at surfaces with nodes, which indicates the presence of geniune string infinities.

---

* or more generally the Cauchy-Riemann operator.

– *T. Hollowood:*

Are factorization and one–loop modular invariance sufficient to generate multiloop modular invariance?

– *L. Alvarez–Gaumè:*

That claim has not rigorously proved although there are strong arguments produced in this respect by Seiberg and Witten. A rigorous proof is simplified by the fact that the modular group for an arbitrary compact Riemann surface has as generators the Dehn twists on one– and two–loop subgraphs. Thus by factorization, it is enough to check that modular invariance holds on these subgraphs. For the bosonic strings, this can probably be checked directly using explicit expressions for two–loop amplitudes. But for fermionic strings the situation is more complicated due to the lack of useful expressions for two–loop amplitudes.

– *J. Quackenbush:*

Do the methods you have presented for studying multiloop closed string amplitudes generalize to open string theories?

– *L. Alvarez–Gaumè:*

Open string diagrams are more complicated because they have boundaries as well as handles . Unfortunately the moduli space for open strings has no complex structure, so the methods of algebraic geometry which were so powerful in dealing with closed strings no longer apply. Calculations must be done using real analytic techniques. But, in fact a fair amount is known about the moduli space for open strings. For instance, Thurston showed how to triangulate it, and Giddings, Martinec and Witten recently showed using Witten's field theory of open strings that the naive Feynman rules of this theory reproduce Thurston's triangulation. The big disadvantage is again, that there is no complex structure and the methods of complex algebraic geometry do not apply.

– *K. Meissner:*

Do additional anomalies arise for strings propagating in a curved background?

– *L. Alvarez–Gaumè:*

Local gauge or gravitational anomalies are local effects, so they will be essentially the same on a curved background space–time. For a background with complicated topology, you may have an interesting group of discontinuous diffeomorphisms, and a rich set of global anomalies. Again, it is standard lore, and strongly believed, that when the sigma model describing the motion of strings in a background is modular invariant, then both the local and global anomalies cancel. Witten showed using some mild topological asuumptions, that there are no global anomalies of any type in the field theory limit of the string on a given background. But the direct connection of this result with string theory is still not completely understood.

– *F. Quevedo:*

It seems that there are at least three acceptable heterotic string theories in 10 dimensions. Those with gauge groups $E_8 \times E_8$, spin (32)/z, and $0(16) \times 0(16)$. Is it possible that these are all just different manifestations of one theory?

– *L. Alvarez-Gaumè:*

The $0(16) \times 0(16)$ theory has infinities due to its nonvanishing cosmological constant so it is actually more a toy than a consistent theory. We do not know how to shift the dilaton tadpole. You can obtain it from the two supersymmetric theories using orbifold–type twists.

A string field theory can have many different ground states. Since we don't know how to compute non–perturbative effects, it is not clear what the potentials for these string fields look like. Without a better understanding of the equations of motion, it is premature to say that these first–quantized string theories are different.

– *M. Yu:*

If the string perturbation expansion breaks down due to strong coupling, what conclusions can be drawn about anomalies in the resulting low–energy effective theory?

– *L. Alvarez-Gaumè:*

We are not even sure how to deal with string theory from a perturbative point of view, so it is not clear to me how to describe a strongly interacting string theory. But the situation is probably similar to what happens in field theory. For example, consider QCD. Then in general, the anomalies you find for such a theory at the fundamental level will be exactly the same as the anomalies in the effective theory of composites beyond the confinement scale. This is a consequence of the 't Hooft anomaly matching conditions. More specifically, remember that $SU(3) \times SU(2) \times U(1)$ is anomaly free when both quarks and leptons are included. But beyond the confinement scale, we have pions instead of quarks. The effective lagrangian that describes the pions is not the naive sigma model – it contains a gauged Wess–Zumino term which makes the theory anomaly free. This term is the only memory the σ–model has that it came from an anomaly free underlying theory. In general once a field theory is formulated for high energies in a consistent, invariant way; at low energies, the theory must remain anomaly free. But of course, the way you realize the cancellation of anomalies may be completely different at strong couplings and low energies than what you might naively expect from the high energy theory. Exactly what mechanism might produce anomaly cancellation in strongly coupled string is something I don't know.

"INOS" AND "SPARTICLES"

Pierre Fayet
Laboratoire de Physique Théorique de l'Ecole Normale Supérieure
24 rue Lhomond, Paris , France

ABSTRACT

We first describe N=1 supersymmetric theories, and present experimental limits on the new particles predicted. We then discuss N=2 extended supersymmetric grand-unified theories, the relations they provide between gauge and Higgs bosons, and how they can be formulated in a 6-dimensional spacetime.

Ten years ago, when I talked about spin-0 leptons and quarks, spin-1/2 gluons and photons, etc., I was considered a mad theoretician, and the idea of a "superworld" seemed totally crazy. By now I am supposed to be a phenomenologist -although no new physics has shown up yet. But these days anyone who lives ordinarily in (3+1) spacetime dimensions, considers pointlike particles and cares about objects which do exist (like photons and electrons, $W^{\pm}$'s and Z's, etc.) is labelled a phenomenologist. Despite the lack of experimental evidence, yesterday's crazy theoretical ideas have become what is often viewed as today's "phenomenology". This is what I will describe in sections I and II of the present lectures.

Beyond that, in section III, I will also discuss the more speculative ideas of extended supersymmetry, which lead us to postulate, again, the existence of more new particles. Such theories may be formulated in an extended spacetime involving additional dimensions of very small sizes. They provide, in particular, a simple physical interpretation for the grand-unification mass, which originates -in four dimensions- from the large values of the momenta carried by the GUT particles along extra space dimensions.

I. N=1 SUPERSYMMETRIC THEORIES OF PARTICLES

Supersymmetric theories are invariant under a set of transformations which change the spins of particles by 1/2 unit, turning bosons into fermions, and conversely. These supersymmetry transformations are generated by a self-conjugate 4-component Majorana spin-1/2 operator Q, which satisfies the algebra[1]

$$\begin{aligned} [Q\ ,\ P^{\mu}] &= 0 \\ \{Q\ ,\ \bar{Q}\} &= -2\ \gamma_{\mu}\, P^{\mu} \end{aligned} \tag{1}$$

(For review articles on supersymmetry and supergravity, see e.g. Refs.2-9.)

For many years supersymmetry had been considered by the vast majority of physicists as a beautiful mathematical structure, very interesting from the point of view of quantum field theory, but of no relevance to particle physics : one could not exhibit even a single pair of known particles of different spins that could be directly related under supersymmetry. The supersymmetry algebra (1) appeared, formally, as a possible extension for the Lorentz and Poincaré algebras underlying the theory of relativity - but the Laws of Nature did not seem to be supersymmetric !

However Nature may be supersymmetric, if one postulates the existence of a whole class of new particles — gluinos and photinos, winos and zinos, spin-0 leptons and quarks, etc. — as yet unobserved, which would be the superpartners of the ordinary particles[10]. Before discussing the properties of the many new particles predicted, it is useful to recall briefly the reasons for introducing them.

1.1 Why Do We Need To Introduce Superpartners ?

Could one use supersymmetry to relate the spin-1 photon with a massless spin-1/2 neutrino — and, at the same time, their charged electroweak partners, the spin-1 $W^{\pm}$'s and the spin-1/2 electrons $e^{\pm}$? This is possible in principle[11], but leads to several difficulties when the other particles are considered. We have to take the other leptons (ν_{μ} , μ^{-} ; ν_{τ} , τ^{-}), and the quarks into account. Of course one could attempt to use N=2 or even N=4 supersymmetry to relate the photon with 2 or even 4 neutrinos (this was actually one of the initial motivations for N=2 supersymmetry[12]). But, even in this case, we would still need new bosonic partners for quarks (these partners cannot be the eight neutral spin-1 gluons). Preserving a symmetry of treatment between leptons and quarks leads one to associate all of them with new bosonic partners.

There are other reasons for not associating the W^- with the electron e^-. This would require part of the Dirac field of the electron to transform, like the W^-, as the lower member of an electroweak triplet (I_3 = -1). We know that this cannot be the case : e^-_L and e^-_R should have I_3 = - 1/2 and 0, respectively.

In the same spirit we might still attempt to relate a left-handed fermionic doublet $\begin{pmatrix} \nu \\ \ell^- \end{pmatrix}_L$ with a doublet of spin-0 fields $\begin{pmatrix} \varphi^{\circ} \\ \varphi^- \end{pmatrix}$, which would trigger the electroweak breaking. But the translation of φ° would result in a charged Dirac field $\ell^-_L + \lambda^-_R$ (λ^- denoting what we now call a gaugino, with I_3 = -1) [11]. Such a field, even if light, cannot, nowadays, be interpreted as describing a charged lepton e^- , μ^- or τ^-. Moreover the field ν_L , originally intended to represent a massless or light neutrino, would acquire a mass by combining with the gaugino λ_{ZR} associated with the Z. Actually,this is how the particles now called "winos" and "zinos" — both mixtures of "gauginos" and "higgsinos" — appeared in supersymmetric electroweak theories. The fermionic partner of the photon (the gaugino field λ_{γ}) which could not finally be identified with any of the known neutrinos, was called "photonic neutrino", subsequently shortened into "photino". Similarly, the spin-1/2 partners of the gluons were called "gluinos", etc.[10].

If supersymmetry relates spin-1 with spin-1/2 particles, or spin-1/2 with spin-0 particles, it does not relate directly any of the known particles together. This was the first apparent failure of supersymmetry. Despite that, Nature may still be supersymmetric, if we postulate the existence of new particles, as yet unobserved, which would be the superpartners of the ordinary ones.

We just mentioned the winos, zinos, photinos and gluinos, which are the spin-1/2 superpartners of the $W^{\pm}$, Z, γ and gluons, respectively. But what about the superpartners of leptons and quarks ?

Relating leptons and quarks with new spin-1 particles (still an open possibility, especially in the framework of extended supersymmetry) would require a very large gauge group ; and, therefore, a large number of new gauge bosons, presumably very heavy, as well as many additional spin-0 bosons, associated with the spontaneous breaking of this large gauge group. The simplest and most economic possibility, which does not require a very large extension of the gauge group, consists in relating all leptons and quarks with new spin-0 particles, often called "sleptons" and "squarks".

The hypothesis of the existence of these new particles was taken more and more seriously, due to the remarkable properties of supersymmetric theories, briefly summarized below :

i) the relations they provide between massive gauge bosons and Higgs bosons ;

ii) their relations with gravitation (supergravity) ;

iii) the improved convergence properties of supersymmetric quantum field theories, leading in particular to the possibility of a solution for the hierarchy problem ;

iv) and, finally, their relations with the fashionable superstring theories.

1.2 Gauge Boson / Higgs Boson Unification in Supersymmetric Theories

Since every known particle gets associated with a superpartner, one could think that making a theory supersymmetric simply consists in duplicating all particles, by an appropriate use of the prefix "s" or the suffix "ino". If suitably constructed, however, supersymmetric theories can provide us with new relations between two classes of particles which are already present in ordinary gauge theories, namely massive spin-1 gauge bosons and spin-0 Higgs bosons. The latter appear (together of course with spin-1/2 inos) as spin-0 partners of spin-1 gauge bosons, in massive multiplets of supersymmetry[10,11,13]. This has important consequences in simple N=1 supersymmetric theories, and even more in extended supersymmetric theories, especially when grand-unification is considered (cf. section III).

A spontaneous breaking of the electroweak SU(2)xU(1) gauge symmetry is required to generate masses for the $W^{\pm}$ and Z bosons by the Englert-Brout-Higgs-Kibble mechanism. With a suitable choice of chiral Higgs superfields one can get a spontaneous breaking of gauge invariance in a supersymmetric theory. Let us describe what happens.

For each spontaneously broken generator of the gauge group, a gauge boson acquires a mass while the corresponding Goldstone boson is eliminated. These two bosons belong to two originally different representations of supersymmetry, which join to give a single massive representation. It is the massive gauge multiplet of supersymmetry describing 1 vector, 1 Dirac (or 2 Majorana) spinor and 1 real spin-0 particle. The latter is precisely a Higgs boson ; cf. Table 1.

Let us illustrate this by considering a U(1) gauge invariant toy model, describing the interactions of an abelian gauge superfield V with a massless left-handed chiral superfield S [14]. The Lagrangian density reads

$$\mathcal{L} = \mathcal{L}_0(V) + \left[S^*(\exp 2eV)S + \xi V\right]_{D\text{-component}} \qquad (2)$$

in which $\mathcal{L}_0(V)$ denotes the free kinetic energy term of V. Let (V^μ, λ) and (ψ_L, φ) be the spin-1 /spin-1/2 and spin-1/2 /spin-0 physical fields described by the superfields V and S, respectively. The Lagrangian density can be expanded as follows :

$$\mathcal{L} = -\frac{1}{4} V_{\mu\nu} V^{\mu\nu} - \frac{i}{2} \bar{\lambda} \not\partial \lambda - i \bar{\psi}_L \not{D} \psi_L - (D_\mu \varphi)^\dagger D^\mu \varphi + i e \sqrt{2} \left[\bar{\psi}_L \lambda \varphi + \varphi^\dagger \bar{\lambda} \psi_L \right] + \frac{D^2}{2} + \xi D + e D \varphi^\dagger \varphi + \frac{F^2 + G^2}{2} \tag{3}$$

After elimination of the auxiliary fields D, F and G, the potential of the scalar field φ reads :

$$V(\varphi) = \frac{D^2}{2} = \frac{1}{2} \left| \xi + e \varphi^\dagger \varphi \right|^2 \tag{4}$$

Table 1 . How the Higgs mechanism operates in supersymmetric theories.

Englert-Brout-Higgs-Kibble mechanism :

1 massless spin-1 gauge boson
+ 1 massless spin-0 Goldstone boson

$\Longrightarrow$ 1 massive spin-1 gauge boson

In a supersymmetric theory :

gauge boson $\in$ massless gauge multiplet
+ Goldstone boson $\in$ massless chiral multiplet

$\Longrightarrow$ Massive gauge boson $\in$ Massive gauge multiplet

with { 1 spin-1 gauge boson
2 Majorana fermions
(or 1 Dirac)
1 spin-0 Higgs boson } all massive

If $\xi e < 0$, φ acquires a non-vanishing expectation value

$$\langle \varphi \rangle = - \frac{v}{\sqrt{2}} \tag{5}$$

with

$$\xi + \frac{1}{2} e v^2 = 0 \tag{6}$$

and supersymmetry remains conserved. The translation (5) generates in the Lagrangian density (3) the following mass terms

$$\mathcal{L}_m = - \frac{1}{2} e^2 v^2 W_\mu W^\mu - i e v \left(\overline{\psi_L} \lambda_R + \overline{\lambda_R} \psi_L \right) - \frac{1}{2} e^2 v^2 \left[\sqrt{2} \, \mathcal{R}e \, \varphi \right]^2 . \tag{7}$$

It shows that

$$\left\{ \begin{array}{ll} \text{- the spin-1 gauge field} & W^\mu \\ \text{- the spin-1/2 Dirac fermion field} & \psi_L + \lambda_R \\ \text{- the spin-0 Higgs boson field} & \sqrt{2} \, \mathcal{R}e \, \varphi \end{array} \right. \tag{8}$$

all acquire the same mass

$$m = e v \ . \tag{9}$$

More generally, massive gauge bosons and Higgs bosons now belong to the same multiplets of supersymmetry. But they are usually described by different superfields, namely, gauge and chiral superfields. Their spin-1/2 partners appear, up to now, as mixings of fields such as λ_R and ψ_L , called gaugino and higgsino fields, respectively.

Alternately, to make the important relation existing between massive gauge bosons and Higgs bosons explicit, we can use <u>massive gauge superfields</u> to describe Higgs bosons. The spin-1/2 components of these superfields will describe both the fermionic fields previously called gauginos and higgsinos.

Let us illustrate this in the simple case of a U(1) gauge model. Instead of choosing the Wess-Zumino gauge [15] to evaluate the Lagrangian density (2) (which becomes polynomial), one can choose another gauge, which preserves manifest supersymmetry, but in which the Lagrangian density remains non-polynomial[14]. This is done by noting that the chiral superfield S can be totally gauged away by means of a generalized gauge transformation. We define the corresponding gauge superfunction Λ by

$$S = - \frac{i v}{2} \exp^{2ie\Lambda} \tag{10}$$

and the new gauge superfield W by

$$W = V + i \left(\Lambda - \Lambda^* \right) . \tag{11}$$

The Lagrangian density (2) is now expressed in terms of the single gauge superfield W :

$$\mathcal{L} = \mathcal{L}_0 (W) + \left[\frac{v^2}{4} \exp^{2eW} + \xi W \right]_D \tag{12}$$

Let us perform a second-order expansion of eq.(12). No linear term appears, owing to eq.(6), and we find

$$\mathcal{L} = \mathcal{L}_0 (W) + \frac{1}{2} e^2 v^2 \left[W^2 \right]_D + \cdots \tag{13}$$

The second term in (13) is precisely the mass term for the massive gauge superfield W, as we now show explicitly.

The real superfield W ($x, \theta, \bar{\theta}$) may be expanded as follows :

$$W(x, \theta, \bar{\theta}) = C + i \theta^\alpha \chi_\alpha - i \bar{\theta}_{\dot{\alpha}} \bar{\chi}^{\dot{\alpha}} - \theta \sigma_\mu \bar{\theta} V^\mu$$
$$+ \ldots$$

.../...

$$+ \frac{i}{2} \theta^\alpha \theta_\alpha (M + iN) - \frac{i}{2} \bar\theta_{\dot\alpha} \bar\theta^{\dot\alpha} (M - iN)$$

$$+ i \theta^\alpha \theta_\alpha \bar\theta_{\dot\beta} \left(\bar\lambda - \frac{i}{2} \partial_\mu \chi \sigma^\mu \right)^{\dot\beta}$$

$$- i \bar\theta_{\dot\alpha} \bar\theta^{\dot\alpha} \theta^\beta \left(\lambda + \frac{i}{2} \sigma^\mu \partial_\mu \bar\chi \right)_\beta$$

$$+ \frac{1}{2} \theta^\alpha \theta_\alpha \bar\theta_{\dot\beta} \bar\theta^{\dot\beta} \left(D + \frac{1}{2} \Box C \right) \tag{14}$$

A second order expansion of eq.(13) gives (with $m = ev$) :

$$\mathcal{L} = -\frac{1}{4} W_{\mu\nu} W^{\mu\nu} - \frac{i}{2} \bar\lambda \not\partial \lambda + \frac{D^2}{2}$$
$$+ \frac{1}{2} m^2 \left[2CD - \partial_\mu C \partial^\mu C - W_\mu W^\mu + M^2 + N^2 - 2i \bar\chi \lambda - i \bar\chi \not\partial \chi \right]$$
$$+ \quad \cdots \tag{15}$$

or, after elimination of the auxiliary fields M, N and D :

$$\mathcal{L} = -\frac{1}{4} W_{\mu\nu} W^{\mu\nu} - \frac{i}{2} \bar\lambda \not\partial \lambda - \frac{i}{2} (\overline{m\chi}) \not\partial (m\chi) - \frac{1}{2} \partial_\mu (mC) \partial^\mu (mC)$$
$$- \frac{m^2}{2} W_\mu W^\mu - i m (\overline{m\chi}) \lambda - \frac{1}{2} m^2 (mC)^2$$

+ non-polynomial interaction terms (16)

It shows that eq.(15) describes a spin-1 gauge field W^μ , 2 spin-1/2 fields λ and ($m\chi$), and a spin-0 Higgs field (mC). All of them have the same mass $m = ev$.

In the new gauge, the spin-0 Higgs field is described, not by a chiral superfield as usual, but by the first (C -) component in the expansion (14) of the massive gauge superfield W. The spinor mass term $-im(\overline{m\chi})\lambda$ may be rewritten as follows :

$$-im(\overline{m\chi})\lambda = -i\frac{m}{2} \left[\overline{\frac{\lambda + m\chi}{\sqrt{2}}} \frac{\lambda + m\chi}{\sqrt{2}} + \gamma_5 \overline{\frac{\lambda - m\chi}{\sqrt{2}}} \gamma_5 \frac{\lambda - m\chi}{\sqrt{2}} \right] . \tag{17}$$

The massive spin-1 gauge boson is associated with a set of two Majorana spin-1/2 inos

$$\frac{\lambda + m\chi}{\sqrt{2}} \qquad \text{and} \qquad \gamma_5 \, \frac{\lambda - m\chi}{\sqrt{2}} \quad . \tag{18}$$

1.3 The Supersymmetric Standard Model

We can now apply the above method to the spontaneous breaking of the electroweak gauge symmetry, and construct the massive $W^{\pm}$ and Z multiplets[10,11,13]. The Z multiplet describes the Z boson, 2 Majorana zinos, and a real spin-0 Higgs boson called z. The $W^{\pm}$ multiplet describes the $W^{\pm}$ boson, 2 Dirac winos and a charged spin-0 Higgs boson called $w^{\pm}$.

To make this construction we need more than one SU(2) doublet Higgs superfield. If we considered the SU(2) and U(1) gauge superfields $\vec{V}$ and V', together with a single doublet Higgs superfield $S = \begin{pmatrix} S^{\circ} \\ S^{-} \end{pmatrix}$, only, we would describe **1 ½** charged Dirac spinor only, namely

$$\lambda^{-} \qquad \text{and} \qquad \psi^{-}_{L} \quad . \tag{19}$$

We do not have enough degrees of freedom to construct our 2 charged Dirac winos ! If no additional superfield is added, we get stuck with an unwanted massless charged fermion.

Therefore we shall supplement (19) with a right-handed charged spinor field ψ^{-}_{R} ,i.e. we shall consider 2 SU(2) doublet chiral Higgs superfields

$$S = \begin{pmatrix} S^{\circ} \\ S^{-} \end{pmatrix} \text{ left-handed} \quad , \quad T = \begin{pmatrix} T^{\circ} \\ T^{-} \end{pmatrix} \text{ right-handed} \tag{20}$$

instead of a single one.

Moreover, these 2 doublet Higgs superfields allow us to generate masses for both charge +2/3 quarks on one hand, and charge -1/3 quarks and charged leptons on the other hand. For example the u and d quarks, which are described by the chiral superfields $Q_L = \begin{pmatrix} U_L \\ D_L \end{pmatrix}$ and U_R , D_R , acquire their masses through the superYukawa couplings

$$\left[2\, h_u \quad T^\dagger\, U_R^\dagger\, Q_L + 2\, h_d \; S^{c\dagger}\, D_R^\dagger\, Q_L \right]_F \quad . \tag{21}$$

When the Higgs superfields S and T acquire vacuum expectation values, which in the simplest case are equal :

$$\langle S^\circ \rangle = \langle T^\circ \rangle = \frac{v}{2} \tag{22}$$

the $W^\pm$ and Z multiplets acquire the masses

$$m_W = \frac{g v}{\sqrt{2}} \quad , \quad m_Z = \frac{(g^2 + g'^2)^{1/2}\, v}{\sqrt{2}} \tag{23}$$

and the u and d quark multiplets the masses

$$m_u = \frac{h_u v}{2} \quad , \quad m_d = \frac{h_d v}{2} \quad . \tag{24}$$

As we said earlier, supersymmetric theories of particles include a charged Higgs boson $w^\pm$ associated with the $W^\pm$, and a neutral one z, associated with the Z. They would be degenerated in mass with the $W^\pm$ and Z, respectively, if supersymmetry were unbroken. After spontaneous breaking of supersymmetry the masses are no longer equal in general, but they still satisfy mass relations, like :

$$\begin{cases} m^2 \text{ (charged Higgs boson } w^\pm \text{)} = m^2_W + \Delta \\ m^2 \text{ (neutral Higgs boson } z \text{)} = m^2_Z + \Delta \end{cases} \quad . \tag{25}$$

Δ depends on the supersymmetry breaking mechanism considered, and is usually positive (e.g. $\Delta = 4\ m^2_{3/2}$) for gravity-induced supersymmetry breaking.

This association between gauge and Higgs bosons is not a trivial property. Gluons are associated with gluinos, which are also color octets. Quarks are associated with spin-0 quarks, which are also color triplets. But the $W^{\pm}$ and Z are associated with Higgs bosons $w^{\pm}$ and z, having different electroweak gauge transformation properties. For example, the $W^{\pm}$ transforms as a member of an electroweak triplet, while the associated Higgs boson $w^{\pm}$ transforms as a member of an electroweak doublet (i.e. $I_3(W^{\pm}) = \pm 1$ differs from $I_3(w^{\pm}) = \pm 1/2$). Moreover the gauge and Higgs bosons have very different couplings to leptons and quarks, proportional to

g and $g \dfrac{m_{\text{fermion}}}{m_W}$, respectively.

Despite that, these gauge and Higgs bosons appear as related under supersymmetry ! Higgs bosons may be described by the lowest spin-0 components in the expansions of the massive gauge superfields $W^{\pm}(x, \theta, \bar{\theta})$, $Z(x, \theta, \bar{\theta})$ - which also describe the spin-1 gauge bosons $W^{\pm}$ and Z, as well as pairs of spin-1/2 inos :

- 2 charged Dirac winos

$$\tilde{W}^{-}_{1,2} = (\gamma_5)\, \frac{\lambda^{-} \pm m_W\, \chi^{-}}{\sqrt{2}} \qquad (\text{mass } m_W) \tag{26}$$

and

- 2 neutral Majorana zinos

$$\tilde{Z}_{1,2} = (\gamma_5)\, \frac{\lambda_Z \pm m_Z\, \chi_Z}{\sqrt{2}} \qquad (\text{mass } m_Z)\,. \tag{27}$$

Winos and zinos are all gaugino/higgsino mixtures, associated under supersymmetry with both the massive spin-1 gauge bosons $W^{\pm}$ and Z, and the massive spin-0 Higgs bosons $w^{\pm}$ and z, in irreducible multiplets of supersymmetry. In the presence of supersymmetry breaking, the mass matrices of winos and zinos should be rediagonalized; there is often one wino lighter than the W, and one zino lighter than the Z, but this is not necessary.

The electroweak symmetry breaking, in a supersymmetric theory, makes use of two doublet chiral Higgs superfields $\begin{pmatrix} S^0 \\ S^- \end{pmatrix}$ and $\begin{pmatrix} T^0 \\ T^- \end{pmatrix}$, left-handed and right-handed, respectively, so as to provide the necessary degrees of freedom for inos and Higgs bosons. The first doublet is also responsible for the masses of charge -1/3 quarks and charged leptons ; the second one for the masses of charge 2/3 quarks. The three chiral superfields S^-, T^- and $\frac{S^0 - T^{0+}}{\sqrt{2}}$ can be gauged away while the gauge superfields $W^{\pm}$ and Z acquire masses. The remaining uneaten neutral chiral Higgs superfield

$$H^0 = \frac{S^0 + T^{0+}}{\sqrt{2}} = \frac{h^0 - i\, h'^0}{2} + \theta\, \tilde{h}^0 + \dots \qquad (28)$$

describes a Majorana spin-1/2 higgsino $\tilde{h}^0$, and two neutral spin-0 Higgs bosons, h^0 (scalar) and h'^0 (pseudoscalar), with dilatonlike and axionlike couplings to leptons and quarks, respectively. Altogether we get, for the supersymmetric extension of the standard model, the following minimal particle content, represented in Table 2 [10].

Table 2 is essentially model-independent, although the mass spectrum of the various particles involved depends on the symmetry breaking mechanisms considered.

In the original model of Ref. 10 supersymmetry was spontaneously broken by means of an extension of the gauge group to SU(3)xSU(2)xU(1)xU(1). This requires the existence of a new neutral gauge boson U, with a predominantly axial coupling to leptons and quarks, related by supersymmetry to the spin-1/2 goldstino generated by the spontaneous breaking of the global supersymmetry. The pseudoscalar h'^0 which appears in

Table 2 is then a would-be Goldstone boson, gauged away while the U boson acquires a mass.

Models of this type, in which an extra U(1) is used to trigger spontaneous supersymmetry breaking, have been very useful as prototypes, allowing one to study the phenomenological properties of the new particles. They are now disfavoured, for several reasons : i) they tend to predict massless or light gluinos, as well as relatively light selectrons $\lesssim$ 40 GeV/c² ; ii) extra Higgs singlets should be introduced, in addition to the two Higgs doublets, in order to avoid unacceptable effects of the extra U(1) gauge boson U ;

Table 2 . Minimal particle content of a N=1 supersymmetric theory, after the spontaneous breaking of the electroweak symmetry. Supersymmetry relates the massive spin-1 gauge bosons $W^\pm$ and Z to the spin-0 Higgs bosons $w^\pm$ and z. (If, in addition, an extra U(1) group is gauged, there is another neutral gauge boson, the U, which acquires a mass while the pseudoscalar Higgs boson h'^0 is gauged away.)

Spin 1	Spin 1/2	Spin 0	
gluons g photon γ	gluinos $\tilde{g}$ photino $\tilde{\gamma}$		
$W^\pm$ Z	2 (Dirac) winos $\tilde{W}_i^\pm$ 2 (Majorana) zinos $\tilde{Z}_i$	$w^\pm$ z	Higgs bosons
	1 (Majorana) higgsino $\tilde{h}^0$	h^0 (standard) h'^0 (pseudoscalar)	
	leptons ℓ quarks q	spin-0 leptons $\tilde{\ell}_L$, $\tilde{\ell}_R$ spin-0 quarks $\tilde{q}_L$, $\tilde{q}_R$	

iii) it is hard to obtain an anomaly-free model while keeping an acceptable mass spectrum ; this requires additional fields, such as mirror lepton and quark fields and additional Higgs doublets, for example.

Nowadays one prefers to use a mechanism of gravity-induced supersymmetry breaking which relies on an "hidden sector" of the supergravity theory (see for example Ref. 5 and references therein). Then the gauge particle of supersymmetry — the spin-3/2 gravitino — acquires a large mass $m_{3/2}$ by the super-Higgs mechanism. One can generate large masses for all spin-0 leptons and quarks at the classical level, without having to introduce an extra U(1) group. When particles coupled with gravitational strength are disregarded, one gets a globally supersymmetric theory with additional soft-breaking[16] terms of dimensions $\leqslant 3$, which are responsible for the mass splittings between bosons and fermions in the multiplets of supersymmetry.

1.4 R-Symmetry And The Interactions Of The New Particles

Let us consider the minimal content of a supersymmetric theory, as given in Table 1. At this stage the theory has an unbroken continuous R-symmetry, defined as follows[10] :

$$V(x, \theta, \bar{\theta}) \longrightarrow V(x, \theta e^{-i\alpha}, \bar{\theta} e^{i\alpha}) \quad (29)$$

for gauge superfields,

$$\begin{cases} S(x, \theta, \bar{\theta}) \ \text{left-handed} \longrightarrow S(x, \theta e^{-i\alpha}, \bar{\theta} e^{i\alpha}) \\ T(x, \theta, \bar{\theta}) \ \text{right-handed} \longrightarrow T(x, \theta e^{-i\alpha}, \bar{\theta} e^{i\alpha}) \end{cases} \quad (30)$$

for the chiral Higgs superfields responsible for gauge symmetry breaking, and

$$\begin{cases} S(x, \theta, \bar{\theta}) \ \text{left-handed} \longrightarrow e^{i\alpha} S(x, \theta e^{-i\alpha}, \bar{\theta} e^{i\alpha}) \\ T(x, \theta, \bar{\theta}) \ \text{right-handed} \longrightarrow e^{-i\alpha} T(x, \theta e^{-i\alpha}, \bar{\theta} e^{i\alpha}) \end{cases} \quad (31)$$

for the chiral superfields describing matter (leptons and quarks).

This corresponds to the conservation of an additive quantum number :

$$\begin{cases} R = 0 & \text{for ordinary particles} \\ & \text{(gauge bosons, Higgs bosons, leptons and quarks)} \\ R = +1, \text{ or } -1 & \text{for their superpartners} \\ & \text{(inos, spin-0 leptons and quarks)} \end{cases} \quad . \tag{32}$$

In particular R-symmetry acts in a chiral way on the self-conjugate Majorana spinors λ representing the gluinos $\tilde{g}$ or the photino $\tilde{\gamma}$:

$$\lambda \rightarrow e^{\gamma_5 \alpha} \lambda \tag{33}$$

Such a continuous R-symmetry, if preserved, will in principle prevent the gluinos and photino from acquiring a mass, even if supersymmetry has been spontaneously broken. This seems experimentally excluded : massless or light gluinos could combine with quarks, antiquarks and gluons, and lead to relatively light hadronic states called R-hadrons[17] ; those would decay into ordinary hadrons by emitting photinos, carrying away part of the energy-momentum. Since no such events have been detected, we need some sort of R-symmetry breaking, so that gluinos can acquire a sizeable mass. The introduction of supergravity - with a massive spin-3/2 gravitino, usually rather heavy - does necessarily lead to violation of this continuous R-symmetry, of the order of the gravitino mass $m_{3/2}$. We shall then get massive gluinos and photinos, either at the tree approximation, or as a result of radiative corrections.

This breaking of the continuous R-symmetry group due to the coupling to supergravity does preserve, however, a discrete subgroup of R-parity transformations. This corresponds to a multiplicatively conserved quantum number called R-parity, $R_p = (-1)^R$. It follows from eqs.(32) that

$$\begin{cases} R_p = + : \text{for ordinary particles} \\ \quad\quad \text{(gauge bosons, Higgs bosons, leptons and quarks)} \\ R_p = - : \text{for their superpartners} \\ \quad\quad \text{(inos, spin-0 leptons and quarks)} \end{cases} \quad (34)$$

This is equivalent to the definition[17] :

$$\text{R-parity} \quad R_p = (-1)^R = (-1)^{(3B + L + 2\,Spin)} \quad (35)$$

The latter formula illustrates that the conservation of baryon and lepton numbers (or simply of their difference B - L) necessarily implies R-parity conservation. A contrario, a breaking of R-parity would tend to generate unwanted B or L violating effects (unless some sort of fine-tuning of the parameters is performed).

Therefore we shall only discuss the "standard phenomenology" of supersymmetry, in which R-parity is conserved. As a result, the new R-odd particles may only be pair-produced. Most of them will be unstable, except the lightest one, which should be absolutely stable. Which is the lightest, stable, R-odd particle depends on the mass spectrum, and is therefore model-dependent. Most of the time one assumes that it is the photino, but it is not necessary. (It might be, also, a spin-0 t or b quark, or a wino or a zino, or a higgsino, etc.).

2. EXPERIMENTAL SEARCHES FOR THE NEW PARTICLES

As we just said, the new particles can only be pair-produced, owing to R-parity conservation. Most of them will be unstable, except the lightest one, here assumed to be the photino. The pair production of superpartners should then lead, ultimately, to the production of a pair of photinos. Photinos have very small interaction cross-sections with matter[18]. Although quasi-invisible, they carry away energy-momentum. "Missing energy" and "missing momentum" are fundamental characteristics of the production of the new particles, and could then be the signature of supersymmetry. This is how most experiments performed to date have been looking for evidence for supersymmetry[4-10].

2.1 Spin-0 Leptons and Photinos

Spin-0 leptons may be pair-produced in e^+e^- annihilations. Assuming the photino to be the "lightest supersymmetric particle", one can search for the process[19]

$$e^+e^- \longrightarrow \tilde{\ell}^+\tilde{\ell}^- \longrightarrow \underbrace{\ell^+\ell^-}_{\text{Non coplanar lepton pair}} + \underbrace{(\tilde{\gamma}\ \tilde{\gamma})}_{\text{2 unobserved photinos}} \qquad (36)$$

which leads to a non-coplanar lepton pair, with at least about 1/2 of the energy missing, in average (see Fig. 1).

Searches performed at PETRA (and also at PEP) have set the following lower bounds on spin-0 lepton masses[20]

$$m(\tilde{e}) > 22\ \mathrm{GeV/c^2}$$
$$m(\tilde{\mu}) > 21\ \mathrm{GeV/c^2}$$
$$m(\tilde{\tau}) > 18\ \mathrm{GeV/c^2} \qquad (37)$$

assuming for simplicity that the two charged spin-0 leptons in

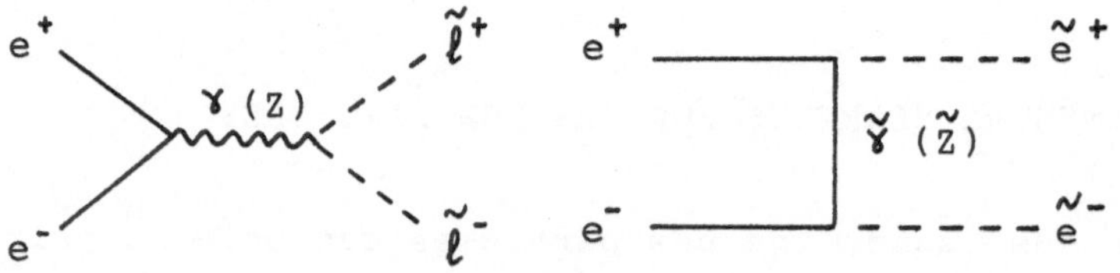

Figure 1 . Pair production of spin-0 leptons in e^+e^- annihilations.

each family are almost degenerated in mass, and that the photino is relatively light. (If $m_{\tilde{\gamma}}$ is close to $m_{\tilde{\ell}}$, the lepton pair produced does not carry much energy and tends to escape detection, so that the constraints get weaker).

The limits (37) are close to the maximum beam energy available at PETRA. But it is also possible to search for the single production of a spin-0 electron, in association with an electron and a photino[19,21] :

$$e^+ e^- \longrightarrow e \quad \tilde{e} \quad \tilde{\gamma} \longrightarrow (e) \ e \ (\tilde{\gamma} \ \tilde{\gamma}) \qquad . \tag{38}$$

This raises the lower limit on selectron masses to nearly 30 GeV/c^2, if the photino is light[22].

Much better limits may be obtained by searching for the radiative production of a pair of photinos, according to the reaction

$$e^+ e^- \longrightarrow \gamma \ (\tilde{\gamma} \tilde{\gamma}) \tag{39}$$

induced by spin-0 electron exchanges[23] (see Fig.2).

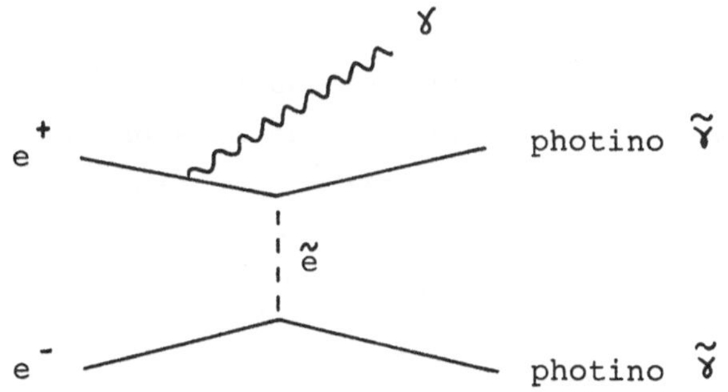

Figure 2 . Radiative production of a photino pair in e^+e^- annihilation.

The experimental signal is the production of a single photon, with missing energy-momentum carried away by the two unobserved photinos. These experiments have now reached such a high degree of precision that one should already start taking into account the concurrent reaction[24] $e^+e^- \longrightarrow \gamma \nu \bar{\nu}$.

Present results from the ASP and MAC experiments at PEP[25] imply

$$m(\tilde{e}) > 65\ \mathrm{GeV}/c^2 \quad (\text{if } m_{\tilde{\gamma}} \lesssim \text{a few GeV}/c^2) \qquad (40)$$

If the photino is heavier, the lower limit on the spin-0 electron mass gets smaller : e.g., for a 10 GeV/c² photino, $m \gtrsim 43\ \mathrm{GeV}/c^2$.

2.2 Spin-0 Quarks and Gluinos

Gluinos and squarks, which are strongly interacting particles, may be copiously produced in strong interaction reactions, like pp or $p\bar{p}$ collisions. They should then decay, according to the reactions :

$$\tilde{q} \longrightarrow q\tilde{g} \ , \quad \tilde{q} \longrightarrow q\tilde{\gamma} \ , \quad \tilde{g} \longrightarrow q\bar{q}\tilde{\gamma} \qquad (41)$$

which ultimately lead to the production of photinos in the final state.

One can search for light gluinos by looking for the reinteraction with matter of the photinos produced by gluino decays, in a beam dump experiment[17,26]. Since the reinteraction cross-section of photinos behaves roughly like[18]

$$\sigma_{(\text{photino + matter})} \sim \frac{\alpha\,\alpha_s}{m_{\tilde{q}}^4}\, s \qquad (42)$$

the limits on the pair production cross-sections of short-lived gluinos — and therefore on their masses — depend on the spin-0 quark masses $m_{\tilde{q}}$. The results of beam dump experiments[27] imply, for example,

$$m_{\tilde{g}} \gtrsim 3 \text{ GeV/c}^2 \quad \text{if} \quad m_{\tilde{q}} < 200 \text{ GeV/c}^2 \tag{43}$$

If spin-0 quarks are heavier, gluinos tend to have longer lifetimes. Lower bounds on their masses - at least of the order of 2 GeV/c² - may then be derived from searches for quasistable particles (see e.g. Ref. 28, and references therein), or from recent experiments searching for the decay $\Upsilon \longrightarrow g\tilde{g}\tilde{g}$ [29,30] at PETRA, or for relatively long-lived R-hadrons that could be produced using a 300 GeV/c π^- beam at the CERN SPS [31].

Another way to search for the new particles is to look for the missing energy, or missing momentum, carried away by the emitted photinos. This method was first used in 1978 to search for light short-lived gluinos in a fixed target experiment performed at Fermilab with a 400 GeV/c proton beam[17]. This experiment[32], performed at a center of mass energy $\sqrt{s} \simeq 27$ GeV, was only sensitive to rather small values of gluino masses $\lesssim 2$ GeV/c². Today, $p\bar{p}$ colliders, which have much higher center of mass energies ($\sqrt{s} \sim 600$ GeV at CERN), provide a much more efficient way to search for the missing energy-momentum that could signal the production of gluinos or squarks, resulting for example from the reaction

$$p\bar{p} \longrightarrow \tilde{g}\tilde{g} \longrightarrow (\tilde{\gamma})q\bar{q} + (\tilde{\gamma})q\bar{q} \tag{44}$$

or from similar processes involving squarks. According to what C. Rubbia told us here, present results from the UA1 experiment imply that gluinos and squarks should probably be heavier than about 100 GeV/c² [33] ; but before using such a figure one should still wait until the analysis of the experimental data gets completed and published.

2.3 Winos and Zinos

Present experiments imply that the superpartners of light particles cannot be too light, i.e. that the "supergap" cannot be too small. This opens the possibility that, conversely, some of the superpartners of heavy particles - like winos and zinos - might be relatively light.

If supersymmetry were unbroken the $W^{\pm}$ would be degenerated in mass with two Dirac spin-1/2 winos, and the Z with two Majorana spin-1/2 zinos, as indicated in Table 2. The actual mass spectrum for these particles — which are mixtures of gaugino and higgsino components — depends on the supersymmetry breaking.

In models with extra U(1) breaking, the two winos are given by

$$\begin{cases} \text{Wino}_1^- = \text{higgsino}_L^- + \text{gaugino}_R^- \\ \text{Wino}_2^- = \text{gaugino}_L^- + \text{higgsino}_R^- \end{cases} \quad (45)$$

They satisfy the mass relation

$$m^2\,(\text{wino}_1) + m^2\,(\text{wino}_2) = 2\,m_W^2 \quad (46)$$

which implies that one wino is lighter than the W, while the other is heavier[10].

This may or may not remain true with gravity-induced supersymmetry breaking, for which the two winos are mixed differently[34,35]. In the absence of direct gaugino mass terms there is always one wino (or zino) lighter than the $W^{\pm}$ (or Z), while the other is heavier. Otherwise, it is also possible that both winos are heavier than the $W^{\pm}$, and both zinos heavier than the Z.

To illustrate this, let us consider as a particular example a simple model with gravity-induced supersymmetry breaking[36], in which the masses of spin-0 leptons and quarks are given by

$$m_{\tilde{l},\tilde{q}} = \left| m_{3/2} \pm m_{l,q} \right| \quad (47)$$

and the masses of the Higgs bosons $w^{\pm}$ and z associated with the $W^{\pm}$ and Z, by

$$\begin{cases} m^2(w^\pm) = m_W^2 + 4\, m_{3/2}^2 \\ m^2(z) = m_Z^2 + 4\, m_{3/2}^2 \end{cases} \tag{48}$$

(up to radiative correction effects).

With the two Higgs doublets acquiring equal vacuum expectation values, we get the following 2x2 mass matrices for winos and zinos (in a higgsino/gaugino basis) :

$$\begin{bmatrix} -m_{3/2} & m_W \\ m_W & m_\lambda \end{bmatrix} , \qquad \begin{bmatrix} -m_{3/2} & m_Z \\ m_Z & m_\lambda \end{bmatrix} , \tag{49}$$

in which m_λ parametrizes direct gravity-induced gaugino mass terms.

In the absence of such terms we find

$$\begin{cases} m\,(\text{winos}) = \left(m_W^2 + \frac{1}{4}\, m_{3/2}^2\right)^{1/2} \pm \frac{1}{2}\, m_{3/2} \\ m\,(\text{zinos}) = \left(m_Z^2 + \frac{1}{4}\, m_{3/2}^2\right)^{1/2} \pm \frac{1}{2}\, m_{3/2} \end{cases} \tag{50}$$

There is necessarily a wino lighter than the W, and a zino lighter than the Z.

On the other hand, with $m_\lambda = -m_{3/2}$, or $+m_{3/2}$, we would get[35]

$$m\,(\text{gluinos}) = m\,(\text{photino}) = m_{3/2} \tag{51}$$

and

$$\begin{cases} m\ (\text{winos}) = |m_W \pm m_{3/2}| \\ m\ (\text{zinos}) = |m_Z \pm m_{3/2}| \end{cases} \tag{52}$$

or, alternately

$$\begin{cases} m\ (\text{winos}) = (m_W^2 + m_{3/2}^2)^{1/2} \\ m\ (\text{zinos}) = (m_Z^2 + m_{3/2}^2)^{1/2} \end{cases} \tag{53}$$

(More precisely, formulas (48,51,53) can be obtained in the framework of N=2 extended supersymmetric theories, as we shall discuss in section III; these formulas will be interpreted in a 5 or 6 dimensional spacetime, in which inos carry momentum $\pm m_{3/2}$, and Higgs bosons $\pm 2 m_{3/2}$, along the extra compact dimensions[37].)

A relatively light wino could be pair-produced in e^+e^- annihilations :

$$e^+e^- \longrightarrow \tilde{W}^+ \tilde{W}^- \tag{54}$$

It could then decay according, for example, to the reactions

$$\begin{cases} \tilde{W}^\pm \longrightarrow \tilde{\gamma}\ \ell^\pm \overset{(-)}{\nu} \\ \tilde{W}^\pm \longrightarrow \tilde{\gamma}\ q\ \bar{q}' \end{cases} \tag{55}$$

induced by $W^\pm$ (or spin-0 lepton or quark) exchanges. Such a wino would look somewhat like an additional charged lepton (see Fig. 3). Present PETRA limits [38] imply that

$$m(\text{winos}) \gtrsim 23 \text{ GeV}/c^2 \tag{56}$$

One may also search for the process

$$e^+e^- \longrightarrow \tilde{Z}\ (\text{unstable})\ \ \tilde{\gamma} \tag{57}$$

induced by spin-0 electron exchanges. This leads to combined constraints on the $\tilde{\gamma}$, $\tilde{Z}$ and $\tilde{e}$ masses[38]. However, in practice, these constraints are not as restrictive as they look : the masses of the various particles are usually correlated in such a way that if $\tilde{\gamma}$ and $\tilde{Z}$ are light enough (so that $e^+e^- \rightarrow \tilde{Z}\tilde{\gamma}$ is kinematically allowed at present energies) the $\tilde{e}$'s are rather heavy, and the expected production cross section (57) is rather small (see for example the mass formulas 47-52).

If sufficiently light, inos could also be produced in $W^\pm$ and Z decays, according for example to the reactions[34,35] :

$$\begin{cases} W^\pm \rightarrow \tilde{W}^\pm\tilde{\gamma}, \ \tilde{W}^\pm\tilde{Z}, \ \tilde{W}^\pm\tilde{h} \\ Z \rightarrow \tilde{W}^+\tilde{W}^-, \ \tilde{Z}\tilde{h}, \ \ldots \end{cases} \qquad (58)$$

They would then decay (see e.g. Fig. 3) :

$$\text{ino} \longrightarrow \text{lighter ino} + q\bar{q}' \ (\text{or } l\bar{l}') \qquad (59)$$

leading to final states with missing energy-momentum, and leptons or quark jets.

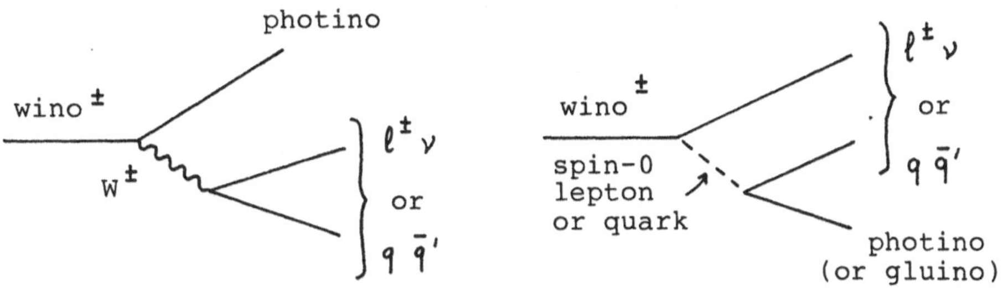

Figure 3 . Possible decay modes of winos.

Collider data imply that the mass m_L of a charged sequential heavy lepton should satisfy[33]

$$m_L \; > \; 41 \text{ GeV}/c^2 \quad . \tag{60}$$

If the photino is light, we have

$$\Gamma(W^\pm \to \tilde{W}^\pm \tilde{\gamma}) \simeq 4 \sin^2\theta \; \Gamma(W^\pm \to L^\pm \nu_L)_{(m_L = m_{\tilde{W}})} \; \frac{m_W^2}{m_W^2 + m_{\tilde{W}}^2} \tag{61}$$

The same collider data should then imply (at least for heavy sleptons and squarks) a lower limit on wino masses nearly as large as for a sequential heavy lepton, i.e.

$$m_{\tilde{W}} \;\gtrsim\; 35 \text{ GeV}/c^2 \tag{62}$$

2.4 A Very Light Gravitino ?

When the supersymmetry algebra is realized locally (supergravity[3]) the spin-2 graviton couples to the energy momentum tensor $T^{\mu\nu}$ while its superpartner, the spin-3/2 gravitino, couples to the conserved supercurrent $J^\mu{}_\alpha$, with strength :

$$\kappa = (8\,\pi\, G_{Newton})^{1/2} \simeq 4.11 \quad 10^{-19} \quad \text{GeV}^{-1} \tag{63}$$

One might think that the production and interaction cross-sections of the gravitino, proportional to G_{Newton}, should be completely negligible. This is indeed true if the gravitino has a sizeable mass — as is usually assumed in most theories with gravity-induced supersymmetry breaking.

However it is not so if the mass of the gravitino, $m_{3/2}$, is very small. The wave-function of an ultrarelativistic gravitino in a ± 1/2 polarization state includes a factor

$$\simeq \sqrt{\frac{2}{3}} \; \frac{k^\mu}{m_{3/2}} \tag{64}$$

so that the corresponding amplitudes, proportional to

$$\frac{\kappa}{m_{3/2}\sqrt{6}} = \frac{1}{d} \tag{65}$$

are not necessarily negligible. As a result, particle physics experiments can provide us with lower limits on the gravitino mass $m_{3/2}$ [39].

Actually a light spin-3/2 gravitino $\tilde{G}$ is produced and interacts like the massless spin-1/2 goldstino of globally supersymmetric theories. The parameter d which appears in eq.(65) has the dimension of a mass², and fixes — by definition — the scale at which supersymmetry is broken, $\sqrt{d} = \Lambda_{SS}$. Supersymmetry may be broken "at a high scale" $\sqrt{d} \gg m_W$, even if boson-fermion mass-splittings remain of order m_W [39]. In that case the goldstino (or gravitino) will be almost invisible in particle physics experiments, as indicated by formula (65). The gravitino mass $m_{3/2}$ and supersymmetry breaking scale $\sqrt{d}$ are related by

$$m_{3/2} \simeq 1.68 \; 10^{-6} \; \text{eV}/c^2 \left(\frac{\sqrt{d}}{100\ \text{GeV}}\right)^2 \tag{66}$$

A first lower limit on $m_{3/2}$ ($> 1.5\ 10^{-8}$ eV/c^2) was obtained from the search for the production of gravitino-photino pairs in ψ decays. Much better limits have been derived now from the searches for

$$\left\{ \begin{array}{l} e^+e^- \longrightarrow \gamma\ \tilde{G} \quad \tilde{\gamma}\ \text{(quasistable)} \\ e^+e^- \longrightarrow \tilde{G} \quad \tilde{\gamma} \\ \qquad\qquad\qquad\quad \hookrightarrow \gamma\ \tilde{G} \end{array} \right. \tag{67}$$

which would both lead to the production of single γ's. From the latest experimental data[22,25] we deduce[40] :

$m_{3/2} > 3\ 10^{-6}\ eV/c^2$

$$\text{or} \quad \sqrt{d} \ (= \Lambda_{SS}) > 140 \text{ GeV} \qquad (68)$$

in the case of light ($\lesssim$ 100 MeV/c^2) quasistable photinos. For heavier — unstable — photinos, the limits are even stronger: $\simeq$ 1.6 10^{-5} eV/c^2 and 300 GeV, respectively.

These limits - which are independent of selectron masses - express the fact that gravitino cross-sections are forbidden to be much larger than ordinary weak interaction cross-sections. Other constraints on $\sqrt{d}$ and $m_{\tilde{\gamma}}$ can also be derived from $e^+e^- \rightarrow 2$ unstable $\tilde{\gamma}$'s, but they depend on the values of the selectron masses, $m_{\tilde{e}}$ [38,41].

III. N=2 SUPERSYMMETRIC GRAND-UNIFIED THEORIES AND EXTRA SPACE DIMENSIONS

3.1 Why N=2 Supersymmetry ?

Simple (N=1) supersymmetric theories cannot be considered as completely satisfactory from a theoretical point of view, especially when grand-unification[42] is introduced. We shall now discuss theories which are invariant under an extended algebra involving N=2 supersymmetry generators Q^i.

What are the motivations for such a construction[37,43-45] ?

i) to realize a much larger association between massive gauge bosons and Higgs bosons, so as to find a place for the many Higgses of supersymmetric GUTs :

$$\begin{array}{ccc} \text{1 massive spin-1} & \text{N=2 supersymmetry} & \text{1 or 5 spin-0} \\ \text{gauge boson} & \longleftrightarrow & \text{Higgs bosons} \end{array} \qquad (69)$$

ii) to reduce the arbitrariness of N=1 theories : the superpotential of N=1 theories becomes fixed by N=2 supersymmetry ; the direct gaugino masses m_λ are also fixed, and equal to the gravitino mass $m_{3/2}$. N=2 supersymmetric GUTs will depend on a very small number of arbitrary parameters, e.g. e, m_X, m_W and $m_{3/2}$, only, for an N=2 SU(5) GUT, at least as long as leptons and quarks are not considered ;

iii) the possibility of determining algebraically the grand-unification mass m_X, from the value of the central charge Z [46] which appears in the N=2 algebra, and is related to the weak hypercharge operator Y [44] :

$$\left\{ Q^i , \bar{Q}^j \right\} = -2 \not{P} \delta^{ij} + 2 Z \varepsilon^{ij} \tag{70}$$

with

$$\left| Z \quad (X^{\pm 4/3}) \right| \equiv m_X \tag{71}$$

iv) the possibility of expressing these theories in a 6 (or maybe 10) dimensional spacetime, in such a way that m_X, $m_{3/2}$ originate from momenta carried by the GUT or SUSY particles along extra compact dimensions, and may be computed, ultimately, in terms of the sizes of these compact dimensions.

N=2 theories have the following general features :

i) existence of larger multiplets and therefore of a new set of gravitinos, photinos, gluinos, winos and zinos, etc.;
ii) existence of spin-0 photons and spin-0 gluons ; and of a spin-1 "graviphoton" ;
iii) existence of mirror leptons and quarks having V+A charged current weak interactions ;
iv) existence of additional relations between spin-1 gauge bosons and spin-0 Higgs bosons.

The motivation for extended supersymmetry is not apparent at this stage ; indeed N=1 supersymmetric theories of weak, electromagnetic and strong interactions, whose minimal content is given in Table 2, are quite appealing. The situation, however, changes when one considers grand-unification : supersymmetric grand-unified theories require a rather large number of spin-0 Higgs bosons.The Higgs sector gets quite complicated, and we no longer have a simple classification, as in Table 1.

Even with only a $\underline{24}$, a $\underline{5}$ and a $\bar{\underline{5}}$ chiral Higgs superfields - certainly a minimal choice for a N=1 supersymmetric SU(5) theory - we get <u>three</u> Higgs bosons of ± 1 unit charge, one of them only being related to the $W^{\pm}$. N=2 supersymmetry will introduce two additional Higgs bosons of ± 1 unit charge, but, ultimately, all <u>five</u> charged Higgses get related with the $W^{\pm}$ [44] :

$$W^{\pm} \longleftrightarrow \text{5 charged Higgs bosons} \qquad (72)$$

and, similarly,

$$Z \longleftrightarrow \text{5 neutral Higgs bosons} \qquad (73)$$

N=2 supersymmetry also leads us to introduce a second octet of gluinos, and a second photino. One may of course question the necessity of introducing them, especially since ordinary gluinos and photinos have not been observed yet. But the second octet of gluinos ($\hat{g}$) and the second photino ($\hat{\gamma}$) are not really new particles. They are already present, although in a hidden way, in a grand-unified theory with only a simple (N=1) supersymmetry. The complex spin-0 Higgs field (such as the $\underline{24}$ of SU(5)) which breaks spontaneously the GUT symmetry describes an octet and a singlet of spin-0 particles, which will be interpreted later as spin-0 gluons and spin-0 photons, respectively. Their fermionic partners are a second octet of Majorana fermions ("paragluinos" $\hat{g}$), similar to the gluinos $\tilde{g}$, and a singlet one ("paraphotino" $\hat{\gamma}$), similar to the photino $\tilde{\gamma}$. Requiring N=2 extended supersymmetry means, in particular, that there is no essential difference between the by-now familiar octet of gluinos, and the second octet of colored fermions which appears as a consequence of grand-unification. Altogether we get two octets of gluinos, two photinos, as well as a complex octet of spin-0 gluons, and a complex spin-0 photon. This is summarized in Fig. 4.

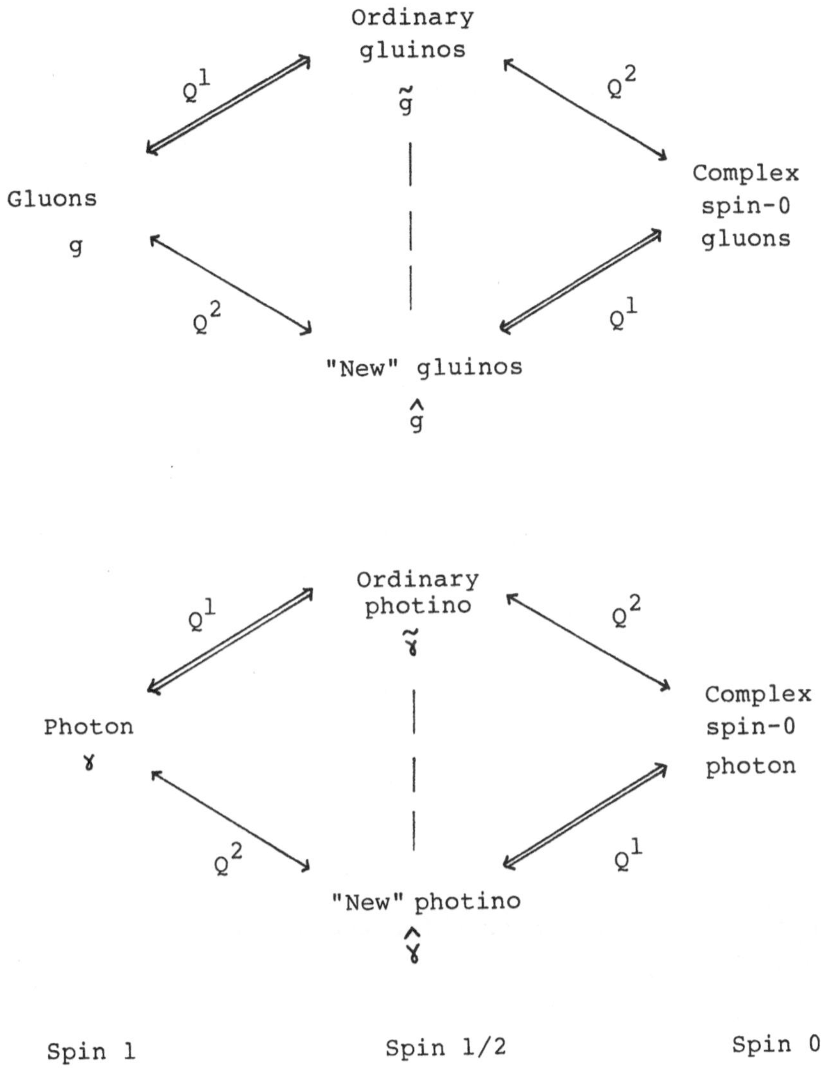

Figure 4 . Relations between the gluons, the photon, and their spin-1/2 and spin-0 partners, in an N=2 extended supersymmetric theory. Q^1 denotes the action of the first supersymmetry generator, Q^2 the action of the second one. Spin-0 gluons and spin-0 photons are described by the complex adjoint Higgs field (e.g. a 24 of SU(5)) which breaks spontaneously the GUT symmetry. (They will appear, subsequently, as the 5th and 6th components of the gluon and photon 6-vector fields, in a 6 dimensional spacetime.)

3.2 The "N=2 Supersymmetric Standard Model"[44]

The extended supersymmetry algebra reads :

$$\left\{Q^1 , \bar{Q}^1\right\} = \left\{Q^2 , \bar{Q}^2\right\} = -2 \not{P} \qquad (74)$$

as in formula (1). In addition, the two supersymmetry generators Q^1 and Q^2 also satisfy an anticommutation relation

$$\left\{Q^1 , \bar{Q}^2\right\} = -\left\{Q^2 , \bar{Q}^1\right\} = 2\, Z \qquad (75)$$

Z is a spin-0 symmetry generator called a central charge[46], which has the dimension of a mass. Its explicit expression includes a spontaneously generated part[43] involving neutral uncolored grand-unification symmetry generators, such as the weak hypercharge $Y = 2(Q - T_3)$.

With SU(5) as the grand-unification group we find :

$$\left\{Q^1 , \bar{Q}^2\right\} = -\left\{Q^2 , \bar{Q}^1\right\} = 2\left[\begin{array}{l}\text{Global symmetry} \\ \text{generator}\end{array} -(3/5)m_X\, Y\right] \qquad (76)$$

The central charge Z vanishes for the $W^{\pm}$ and Z, γ and gluons, but equals

$$Z = \mp\, m_X \qquad (77)$$

for the GUT bosons $X^{\pm 4/3}$ and $Y^{\pm 1/3}$ (which have weak hypercharges $Y = \pm\, 5/3$). We also find the mass relation

$$m_Y^2 = m_X^2 + m_W^2 \qquad (78)$$

In N=2 extended supersymmetry the minimal particle content gets increased compared with that given in Table 2. We give in Table 3 the minimal particle content of a N=2 supersymmetric SU(5) theory, after gauge symmetry breaking. As a consequence of extended supersymmetry, almost every Higgs boson now appears as a superpartner of a gauge boson. (This association becomes complete if an extra U(1) is gauged.)

Note the existence of 3 different types of massive gauge hypermultiplets, with different field contents[43] :

- type I (like the $W^{\pm}$ and Z) : they describe 1 massive gauge boson, 4 spin-1/2 inos and 5 spin-0 Higgs bosons ; and carry no central charge :

$$\mathcal{M} > |Z| = 0 \qquad (79)$$

- type II (like the $X^{\pm 4/3}$) : they describe 1 massive gauge boson, 2 spin-1/2 inos and 1 spin-0 Higgs boson, all complex ; they carry a non-vanishing value of the central charge Z, and verify :

$$\mathcal{M} = |Z| > 0 \qquad (80)$$

- type III (like the $Y^{\pm 1/3}$) : they describe 1 massive gauge boson, 4 spin-1/2 inos and 5 spin-0 Higgs bosons, all complex ; they carry a non-vanishing value of the central charge Z, and verify :

$$\mathcal{M} > |Z| > 0 \qquad (81)$$

The actual mass spectrum of the new particles will depend on the way in which the symmetry breaking is performed. If the supersymmetry breaking is induced by gravitation or by dimensional reduction, one can get the following kind of mass spectrum[37] :

$$m_{gluinos} = m_{photino} = m_{3/2} \qquad (82)$$

$$\begin{cases} m\ (winos) = (m_W^2 + m_{3/2}^2)^{1/2} \\ m\ (zinos) = (m_Z^2 + m_{3/2}^2)^{1/2} \end{cases} \qquad (83)$$

and

Table 3 . Minimal particle content of an N=2 supersymmetric SU(5) grand-unified theory. The spontaneous breaking $SU(5) \rightarrow SU(3)_{QCD} \times U(1)_{QED}$ is induced by a complex adjoint 24, and 4 quintuplets (5 and $\bar{5}$) of spin-0 Higgs fields[44].

Spin 1	Spin 1/2	Spin 0
$X^{\pm 4/3}$ $Y^{\pm 1/3}$	2X3 Dirac xinos 4X3 Dirac yinos	3 charged Higgs bosons 5X3 charged Higgs bosons
$W^{\pm}$ Z photon γ gluons g	4 Dirac winos 4 Majorana zinos 2 Majorana photinos 2X8 Majorana gluinos	5 charged Higgs bosons 5 neutral Higgs bosons 2 spin-0 photons 2X8 Spin-0 gluons
	2 neutral Majorana higgsinos	4 neutral Higgs bosons
	leptons and quarks + mirror partners	spin-0 leptons and quarks + mirror partners
+ gravitation multiplet : 1 spin-2 graviton, 2 spin-3/2 gravitinos, 1 spin-1 "graviphoton"		

$$\begin{cases} m \text{ (charged Higgs bosons)} = m_W, \text{ or } (m_W^2 + 4\, m_{3/2}^2)^{1/2} \\ m \text{ (neutral Higgs bosons)} = m_Z, \text{ or } (m_Z^2 + 4\, m_{3/2}^2)^{1/2} \end{cases} \qquad (84)$$

(up to radiative correction effects), as will be discussed in subsection 3.4. A crucial question is, of course, whether both N=2 supersymmetry generators break simultaneously, or does one have a sequential breaking $N=2 \rightarrow N=1 \rightarrow N=0$? In the second case we would no longer expect mirror leptons and quarks, and spin-0 photons and gluons, to be present at relatively low energies.

3.3 Proton Stability in N=2 Supersymmetry GUTs

A consequence of the appearance of the weak hypercharge operator Y in the anticommutation relation of the two different supersymmetry generators (eq. (76)) is the existence of mass splittings $\simeq m_X$ in all multiplets of the grand-unification group, i.e. in lepton and quark multiplets as well as in the gauge boson multiplet.

As a result the grand-unification symmetry associates light leptons with heavy quarks of mass $\simeq m_X$ (or 2 m_X), and conversely. It is therefore necessary to perform a replication of representations, in order to describe every single family of quarks and leptons[47].

It follows that the $X^{\pm 4/3}$ and $Y^{\pm 1/3}$ gauge bosons (as well as their Higgs partners) do not couple directly light quarks to light leptons. The usual diagrams responsible for the standard proton decay mode $p \rightarrow \pi^0 e^+$ do not exist ! (cf. Fig. 5). Actually in minimal N=2 SUSY GUTs, the proton tends to be totally stable.

3.4 Supersymmetric GUTs In A 6 Dimensional Spacetime

An additional interest of extended supersymmetric theories is that they can be formulated in a 6 dimensional spacetime[37,43-45,48]. The two photinos appearing in Fig. 4, or Table 3, originate from a single Weyl (chiral) spinor in 6 dimensions :

1 Weyl photino in 6 dim.	$\longrightarrow$	2 Majorana photinos in 4 dim.	(85)
1 Weyl gluino octet in 6 dim.	$\longrightarrow$	2 Majorana gluino octets in 4 dim.	(86)

Weyl spinors in 6 dimensions also describe, at the same time, ordinary leptons and quarks as well as their mirror partners.

Figure 5 . Forbidden couplings in N=2 SUSY GUTs : the $X^{\pm 4/3}$ and $Y^{\pm 1/3}$ gauge bosons do not couple directly to light leptons and light quarks.

The $W^{\pm}$ and Z masses are already present in the 6 dimensional spacetime (i.e., they can be generated in a 6 d Poincaré invariant way). In 6 dimensions the $X^{\pm 4/3}$ is still massless, while the $Y^{\pm 1/3}$ has the same mass as the $W^{\pm}$:

$$\begin{cases} m^{6d}\ (X^{\pm 4/3}) = m^{6d}\ (\gamma) = m^{6d}\ (\text{gluons}) = 0 \\ m^{6d}\ (Y^{\pm 1/3}) = m^{6d}\ (W^{\pm}) = m_W \\ m^{6d}\ (Z) = m_Z = \dfrac{m_W}{\cos\theta} \end{cases} \tag{87}$$

This corresponds to the existence, in the 6 dimensional spacetime, of a manifest electrostrong symmetry $SU(4) \supset SU(3)_{QCD} \times U(1)_{QED}$, which relates the photon with the eight colored gluons. It survives the "electroweak breaking" $SU(5) \rightarrow SU(4)$ (or $O(10) \rightarrow SU(4)$, etc.), which generates a mass for the $W^{\pm}$ and Z bosons.

The grand-unification mass in 4 dimensions appears as the result of the dimensional reduction from 6 (or 5) to 4 dimensions. The central charge Z, which is present in the N=2 supersymmetry algebra (76), gets replaced by the fifth component of the covariant momentum

$$Z \rightsquigarrow \mathcal{P}^5 = -i\,\mathcal{D}^5 \tag{88}$$

The grand-unification gauge bosons $X^{\pm 4/3}$ and $Y^{\pm 1/3}$ now carry covariant momentum $|\mathcal{P}^5| = m_X$ along the compact fifth dimension.

The 4-dimensional mass spectrum is given by the general formula :

$$\mathcal{M}^2_{4d} = \mathcal{M}^2_{6d} + (\mathcal{P}^5)^2 + (\mathcal{P}^6)^2 \qquad (89)$$

which implies in particular

$$m_Y^2 = m_W^2 + m_X^2 \qquad (90)$$

thereby providing us with a simple physical interpretation for this N=2 mass formula[44].

In a similar way we can also use the R-invariance of the 6-dimensional theory to generate mass-splittings between bosons and fermions in the 4-dimensional spacetime. The definitions (29-32) of R-invariance may be extended to 6-dimensional theories[37]. The 6d R-quantum numbers of the various fields are as follows :

$$\begin{cases} R = 0 \text{ for gauge bosons} \\ R = \pm 1 \text{ for inos (most of them gaugino/higgsino mixtures)} \\ R = 0, \text{ or } \pm 2 \text{ for Higgs bosons} \end{cases} \qquad (91)$$

$$\begin{cases} R = 0 \quad \text{for leptons and quarks} \\ R = \pm 1 \quad \text{for spin-0 leptons and quarks} \end{cases} \qquad (92)$$

As a result one can arrange in such a way that inos, and spin-0 leptons or quarks, carry momentum $\pm\, m_{3/2}$ along the compact fifth dimension — for example — while Higgs bosons carry momentum 0 or $\pm\, 2\, m_{3/2}$. The mass formula (89) does then lead to the 4d mass spectrum :

$$m\,(\text{photino}) = m\,(\text{gluinos}) = m_{3/2}$$

$$\begin{cases} m^2(\text{winos}) = m_W{}^2 + m_{3/2}{}^2 \\ m^2(\text{zinos}) = m_Z{}^2 + m_{3/2}{}^2 \end{cases}$$

$$\begin{cases} m^2\ (\text{charged Higgs bosons}) = m_W{}^2,\ \text{or}\ (m_W{}^2 + 4\, m_{3/2}{}^2) \\ m^2\ (\text{neutral Higgs bosons}) = m_Z{}^2,\ \text{or}\ (m_Z{}^2 + 4\, m_{3/2}{}^2) \end{cases}$$

$$m_{\tilde{\ell}} \simeq m_{\tilde{q}} \simeq m_{3/2} \qquad \text{etc.} \qquad (93)$$

in which we have disregarded the effects of radiative corrections, and of lepton and quark masses (leptons and quarks still remain massless at this stage).

Up to now m_X and $m_{3/2}$ were arbitrary parameters. One can also relate them, in various possible ways, to the lengths of the extra compact dimensions, L_5 and L_6 [37,44] ; for example :

$$m_X = \frac{\pi \hbar}{L_5\, c}, \qquad m_{3/2} = \frac{\pi \hbar}{L_6\, c} \qquad (94)$$

Although they are probably unrealistically simple (especially since the space formed by the extra compact dimensions is likely to be more complicated than a flat torus), these formulas illustrate how both the grand-unification and supersymmetry breakings may appear, in 4 dimensions, as an effect of extra compact dimensions of spacetime ; and how some of the fundamental scales of particle physics may be computed in terms of geometric parameters associated with this compact space.

In ordinary supersymmetric theories superpartners are usually expected to show up at a mass scale $\sim m_W$, or in any case $\lesssim O$ (TeV/c^2). Extended supersymmetric theories, and higher dimensional theories, however, do also suggest less conventional possibilities. While both m_x and $m_{3/2}$ could be very large - in which case supersymmetry would only become apparent at very high energies - it is also possible that $m_{3/2}$ remains of order $\sim m_W$. In that case a relation like

$$m_{3/2} \sim \frac{\hbar}{Lc}$$

would then indicate that both supersymmetry and extra dimensions might show up at future colliders.

REFERENCES

1 Yu A. Gol'fand and E.P. Likhtman, J.E.T.P. Lett. 13, 323 (1971) ;
D.V. Volkov and V.P. Akulov, Phys. Letters 46B, 109 (1973) ;
J. Wess and B. Zumino, Nucl. Phys. B70, 39 (1974).

2 P. Fayet and S. Ferrara, Phys. Reports 32C, 249 (1977).

3 P. Van Nieuwenhuizen, Phys. Reports 68C, 189 (1981).

4 J. Ellis et al., Phys. Reports 105, 1 (1984).

5 H.P. Nilles, Phys. Reports 110, 1 (1984).

6 P. Fayet, Proc. of the 1984 Trieste Spring School on Supersymmetry and Supergravity (World Scientific, Singapour) 114 (1984).

7 H. Haber and G. Kane, Phys. Reports 117, 75 (1985).

8 M. Sohnius, Phys. Reports 128, 39 (1985).

9 Proc. of the 13th SLAC Institute on Particle Physics (1985).

10 P. Fayet, Phys. Letters 64B, 159 (1976) ; 69B, 489 (1977) ; "Unification of the Fundamental Particle Interactions", Proc. Europhysics Study Conf., Erice, 1980, eds. S. Ferrara, J. Ellis and P. Van Nieuwenhuizen (Plenum, N.Y.), 587 (1980).

11 P. Fayet, Nucl. Phys. B90, 104 (1975).

12 P. Fayet, Nucl. Phys. B113, 135 (1976).

13 P. Fayet, Nucl. Phys. B237, 367 (1984).

14 P. Fayet, Nuovo Cimento 31A, 626 (1976).

15 J. Wess and B. Zumino, Nucl. Phys. B78, 1 (1974).

16 L. Girardello and M. Grisaru, Nucl. Phys. B194, 65 (1982).

17 G.R. Farrar and P. Fayet, Phys. Letters 76B, 575 (1978) ; 79B, 442 (1978).

18 P. Fayet, Phys. Letters 86B, 272 (1979).

19 G.R. Farrar and P. Fayet, Phys. Letters 89B, 191 (1980).

20 CELLO collaboration, Phys. Letters 114B, 287 (1982) ;

JADE collaboration, Phys. Letters 152B, 385 (1985) ; 152B, 392 (1985) ;
MARK J collaboration, Phys. Letters 152B, 439 (1985) ;
TASSO collaboration, Phys. Letters 117B, 365 (1982) ;
See also S.L. Wu, preprint DESY 86-007.

21 M.K. Gaillard, L. Hall and I. Hinchliffe, Phys. Letters 116B, 279 (1982).

22 CELLO collaboration, contribution submitted to the 23rd Int. Conf. on High-Energy Physics at Berkeley (U.S.A.), (1986).

23 P. Fayet, Phys. Letters 117B, 460 (1982) ;
J. Ellis and J.S. Hagelin, Phys. Letters 122B, 303 (1983).

24 E. Ma and J. Okada, Phys. Rev. Lett. 41, 287 (1978) ;
K.J.F. Gaemers, R. Gastmans and F.M. Renard, Phys. Rev. D19, 1605 (1979).

25 E. Fernandez et al., Phys. Rev. Lett. 54, 1118 (1985) ;
G. Bartha et al., Phys. Rev. Lett. 56, 685 (1986) ;
H. Küster, Proc. of the 21st Rencontre de Moriond at Les Arcs (France), (1986) ;
N. Jonker, Proc. of the 21st Rencontre de Moriond at Les Arcs (France), (1986) ;
M. Davier, talk given at the 23rd Int. Conf. on High-Energy Physics at Berkeley (U.S.A.), (1986).

26 G. Kane and J. Léveillé, Phys.Letters 112B, 227 (1982).

27 CHARM collaboration, Phys. Letters 121B, 429 (1983) ;
R.C. Ball et al., Phys. Rev. Lett.53, 1314 (1984) ;
W.A. 66 collaboration, Phys. Letters 160B, 212 (1985).

28 S. Dawson, E. Eichten and C. Quigg, Phys. Rev. D31, 1581 (1985).

29 B.A. Campbell, J. Ellis and S. Rudaz, Nucl.Phys. B198, 1 (1982).

30 ARGUS collaboration, Phys. Letters 167B, 360 (1986).

31 NA3 collaboration, Z. Phys.C31, 21 (1986).

32 J.P. Dishaw et al., Phys. Letters 85B, 142 (1979).

33 C. Rubbia, these Proceedings.

34 S. Weinberg, Phys. Rev. Lett. 50, 387 (1983) ;
R. Arnowitt, A.H. Chamseddine and P. Nath, Phys. Rev. Lett. 50, 232 (1983).

35 P. Fayet, Phys. Letters 125B, 178 (1983) ; 133B, 363 (1983).

36 E. Cremmer, P. Fayet and L. Girardello, Phys. Letters 122B, 41 (1983).

37 P. Fayet, Nucl. Phys. B263, 649 (1986), in particular the Appendix.
38 S. Komamiya, Proc. of the 1985 Int. Symposium on Lepton and Photon Interactions at High Energies (Kyoto, Japan) 611 ;
S.L. Wu, preprint DESY 86-007 ;
H. Küster, Proc. of the 21st Rencontre de Moriond at Les Arcs (France), (1986) ; and references therein.
39 P. Fayet, Phys. Letters 70B, 461 (1977) ; 84B, 421 (1979).
40 P. Fayet, Phys. Letters 175B, 471 (1986).
41 CELLO collaboration, Phys. Letters 123B, 127 (1983) ;
JADE collaboration, Phys. Letters 139B, 327 (1984) ;
MARK J collaboration, MIT-LNS Reports 139/1984 ;
TASSO collaboration, Z. Physics C26, 337 (1984).
42 H. Georgi and S.L. Glashow, Phys. Rev. Lett. 32, 438 (1974) ;
H. Georgi, H.R. Quinn and S. Weinberg, Phys. Rev. Lett. 33, 451 (1974) ;
P. Langacker, Phys. Reports 72, 185 (1981) ; and references therein.
43 P. Fayet, Nucl. Phys. B149, 137 (1979).
44 P. Fayet, Nucl. Phys. B246, 89 (1984) ; Phys. Letters 146B, 41 (1984).
45 P. Fayet, Phys. Letters 159B, 121 (1985).
46 R. Haag, J.T. Łopuszański and M. Sohnius, Nucl. Phys. B88, 257 (1975).
47 P. Fayet, Phys. Letters 153B, 397 (1985).
48 F. Gliozzi, J. Scherk and D. Olive, Nucl. Phys. B122, 253 (1977) ;
L. Brink, J. Scherk and J.H. Schwarz, Nucl. Phys.B121, 77 (1977).

DISCUSSION

CHAIRMAN: P. Fayet

Scientific Secretaries: Z. Bern, D. Cocolicchio, G. D'Ambrosio and S. Webb

– *S. Carlip:*

Why do we see electrons and no selectrons; photons and no photinos? Is there any model–independent way which tells you which particles in a supermultiplet stay light?

– *P. Fayet:*

We construct our theories this way because that is what we see. The simplest theories would have unbroken supersymmetry.

– *M. McGuigan:*

Could you compare the phenomenology of N=2 supersymmetry with that obtained from N=1?

– *P. Fayet:*

N=2 supersymmetry theories are not realistic yet because they are too much constrained and we do not know how to get a satisfactory mass spectrum for leptons and quarks. If we forget about that, the phenomenology of N=2 theories will be rather similar to that of N=1 theories. There will be as usual squarks and sleptons, but also extra gluinos and photinos, spin–0 gluons and spin–0 photons and extra Higgs bosons. There should also be mirror leptons and quarks with V + A charged current weak interactions. Whether or not they should actually be present at relatively low energies may depend on how the N=2 supersymmetries get broken.

– *G. Coughlan:*

In order to keep gauge symmetry breaking separate from supersymmetry breaking, presumably one needs independent sizes for the 5th and 6th dimensions. Does this mean that one should consider compactification on a torus?

– *P. Fayet:*

Not necessarily. In fact five dimensions can be sufficient to obtain both gauge and supersymmetry breakings, the scales of which are not necessarily equal. One should, however, worry about the possible existence of mirror fermions.

– *G. D'Ambrosio:*

I would like to speak about gluino masses. There are two bounds: one from beam–dump experiments (low energy) and another from $p\bar{p}$ (high energy). Is there a window between these two?

– *P. Fayet:*

According to C. Rubbia such a window is now closed, but you should still wait for the final results of the analysis.

– *M. Quiros:*

In local SUSY theories the gravitino mass is sometimes decoupled from the scale of SUSY breaking in the observable sector. In that case $m_{\frac{3}{2}} << M_{Planck}$. Is there a lower bound on $m_{\frac{3}{2}}$? and if so where does it come from?

– *P. Fayet:*

The lower bound of gravitino mass, $m_{\frac{3}{2}} > 3 \times 10^{-6}$ eV, can be obtained from the amplitudes

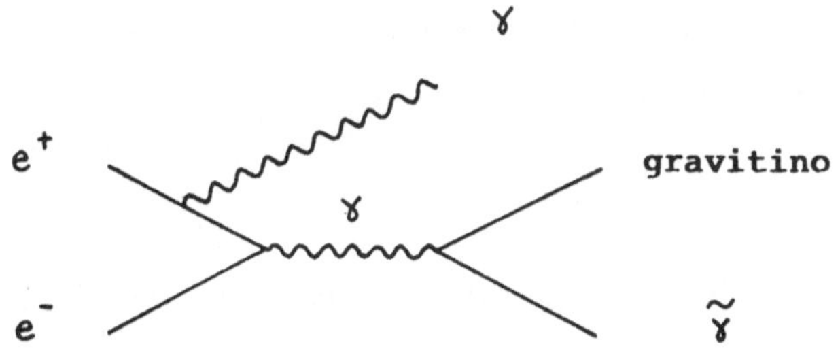

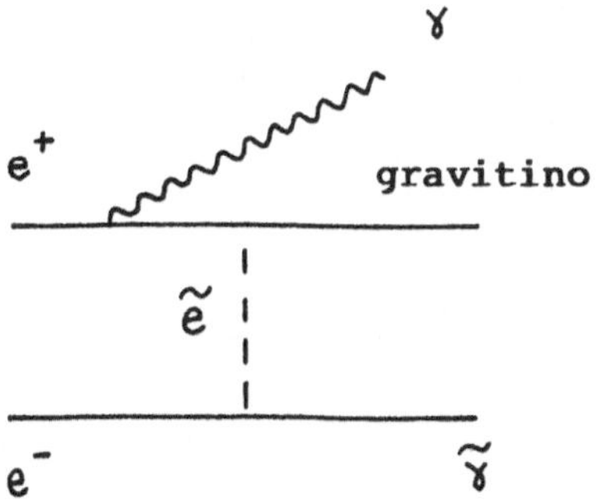

,

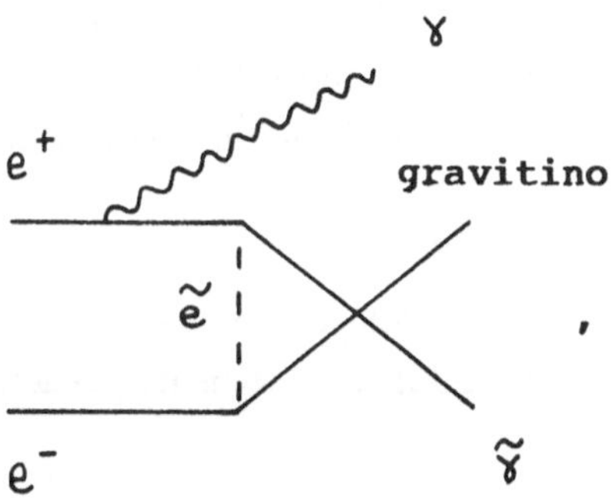

, etc.,

which would be responsible for the process $e^+e^- \to$ single γ + unobserved ($\tilde{\gamma}$ + gravitino). If the gravitino has not–too–small mass its production and interaction cross-sections will be negligible because they are proportional to the Newton gravitation constant G_N; the direct effects of gravitinos are negligible in particle physics.

On the other hand, if the gravitino is extremely light the production and interaction cross–section of gravitinos of the polarization $\pm\,\frac{1}{2}$, proportional to $G_N/m^2_{\frac{3}{2}}$, are not necessarily negligible if the gravitino mass is sufficiently small. From this we derived the previous lower bound for the gravitino mass; or equivalently we can put a constraint on the scale of SUSY breaking which should be larger than 140 GeV. This bound is derived in the case of a relatively light photino. If the mass of the photino is larger than about 100 MeV, we also have to consider the decy $\tilde{\gamma} \to \gamma$ + gravitino, which makes the photino unstable.
From the study of the process

$$e^+e^- \to \text{gravitino} + \tilde{\gamma}$$

$$\hookrightarrow \gamma + \text{ gravitino}$$

we get a more restrictive bound

$$m_{\frac{3}{2}} > 10^{-5}\ eV$$

provided that the mass of the photino is between about 200–300 MeV and 10–15 GeV.

– *M. Quiros:*

One or two years ago there was some controversy about the possibility of a sequential breaking of N=2 supergravity. What is the status of this problem?

– *P. Fayet:*

There are indeed models where the sequential breaking N=2 supergravity $\to$ N=1 $\to$ N=0 is performed. However no application of this sequential breaking to N=2 theories of particles has been done yet.

– *A. Shapere:*

Is there a Goldstone boson associated with the breakdown of continuous R–invariance, and if so, what is its phenomenology?

– *P. Fayet:*

It depends on the kind of breaking of R–symmetry you consider. If the breaking is spontaneous then you obtain a Goldstone boson associated with this breaking. However, in models with gravity–induced supergravity breaking, the continuous R–invariance is explicitly broken down to the discrete subgroup of R–parity transformations, and no such Goldstone boson appears.

– *Z. Bern:*

Is it possible to construct realistic N=2 models which do not have extra compactified dimensions?

– *P. Fayet:*

As far as I know, N=2 supersymmetry theories are not yet realistic. The reason is that they are too much constrained. In N=1 theory one has an arbitrary superpotential and so you have arbitrary Yukawa coupling constants and therefore you have to put in by hand the masses for leptons and quarks. In N=2 supersymmetry the superpotential is totally fixed by the second supersymmetry invariance and therefore you do not know how to get these arbitrary parameters for the masses.

The best you can do at present is to have quarks and leptons massless on the grand unification scale. This is already difficult to achieve, since the N=2 supersymmetric theories are vector–like and each particle has a mirror particle. They are connected to each other by a $\underline{24}$ adjoint which acquires a large v.e.v. Therefore leptons and quarks have a strong tendency to combine with the mirror partners and acquire very large masses of the order of the grand–unification mass. But this problem can be solved, and one recovers leptons and quarks which remain massless, compared with the grand–unification scale.

– *P. Fisher:*

You showed limits on the slepton masses being $m_{\tilde{e}} > 22$ GeV, $m_{\tilde{\mu}} > 21$ GeV and $m_{\tilde{\tau}} > 18$ GeV. Naively, I would expect either all the limits to be the same or the selectron limit to be lower. Can you explain why this is not so?

– *P. Fayet:*

The reason is that the graphs which are responsible for the pair production of the spin–0 electrons, muons and taus are not the same. In the case of spin=0 electrons you can exchange a photino (in the t–channel) in addition to a photon (in the s–channel) and this photino exchange contribution is relatively large. So one expects a larger rate for the production of spin–0 electrons as compared to spin–0 muons and taus. As a result one gets the following different limits

$$m_{\tilde{e}} > 22 \ GeV$$

$$m_{\tilde{\mu}} < 21 \ GeV$$

$$m_{\tilde{\tau}} > 18 \ GeV$$

The limit in the spin–0 taus is even lower because taus are more difficult to identify than muons.

– *A. Klegge:*

C. Rubbia mentioned the possibility of counting neutrino flavours in $e^+e^- \rightarrow \gamma\nu\bar{\nu}$. How do you distinguish this process from $e^+e^- \rightarrow \gamma\tilde{\gamma}\tilde{\gamma}$?

– *P. Fayet:*

Up to now only few events $e^+e^- \rightarrow$ "1 single γ" have been observed. If you start finding an excess of such events compared with what is expected in the standard model with three flavours, then you can start asking whether the extra events are due to the production of additional neutrino pairs.

At the present stage you can only put an upper limit on the number of neutrinos, and, also a lower limit on the selectron mass of about 65 GeV, provided, of course, the photino is sufficiently light.

– *A. Klegge:*

Are there any speculations about R–parity breaking? If so, what do you think the phenomenological effects of such a breaking would be?

– *P. Fayet:*

The problems with R–parity breaking is that, necessarily, when you have R–parity breaking then at the same time you must have breaking of lepton number conservation, baryon number conservation, or both. We know that at least up to a good approximation, lepton number and baryon number are conserved.

If you want to break R–parity then most of the time you find that it does not work. If you add–on R–parity breaking terms to the Lagrangian density the extra terms tend to induce new processes that you do not want, such as too fast proton decay, masses for neutrinos, and so on.

– *D. Rohrlich:*

Although supersymmetry puts the gauge and Higgs bosons into a special relationship, does it constrain the parameters which are normally required to fix the particle masses, or are those parameters arbitrary in the SUSY version of the theory?

– *P. Fayet:*

That is precisely the reason why, in my opinion, if you start considering GUT's you should also work with extended supersymmetry. In ordinary N=1 supersymmetric theories you have a large number of Higgs fields, and a relatively large number of arbitrary coupling constants which appear in the superpotential. The situation is completely different in N=2 extended supersymmetric theories, where you no longer have any arbitrary parameter in the superpotential. N=2 SUSY depends on few arbitrary parameters: 1) the gauge coupling constant, 2) the grand–unification mass, 3) the W mass, 4) the gravitino mass. So with only 4 arbitrary parameters you have a theory of N=2 SUSY $\times SU(5)$. $SU(5)$ breaks down to $SU(3)\times SU(2)\times U(1)$ which further breaks to $SU(3)\times U(1)$. Two of these parameters, the grand unification mass and the gravitino mass, may be considered as having a geometrical origin: if you buy the higher–dimensional interpretation of N=2 supersymmetric theories, you can compute these masses as proportional to the inverse lengths of the additional compact dimensions.

– *A. Pasquinucci:*

Could you explain the relation between the central charge and the unification mass in N=2 supersymmetric theory?

– *P. Fayet:*

In the framework of N=1 supersymmetric theories there is only a single type of massive gauge multiplet, including a massive spin–1 particle, a massive spin–0 Higgs boson and two Majorana spinors. In N=2 supersymmetric theories, on the other hand, there are 3 types of massive gauge multiplets with different particle contents, as discussed in the lecture. The equality between the mass of the grand–unification gauge boson $X^{\pm\frac{4}{3}}$ and the value of the central charge is related to the reduced content of this multiplet: the X boson is associated with only 2 xinos (instead of 4) and 1 Higgs boson (instead of 5) – all of them being, of course, color triplets.

PROTON DECAY IN THE SUPER-WORLD

Stuart Raby

T-8, MS B285, Los Alamos National Laboratory

Los Alamos, NM 87545

INTRODUCTION

One of the most dramatic predictions of any grand unified theory (GUT) is for the decay of the nucleon.[1] The experimental observation of this decay could illuminate the physical principles operating on scales as small as 10^{-30}cm. In this talk we shall elaborate on the predictions for nucleon decay in supersymmetric (SUSY) GUTs.

General Analysis

The experimental limits on nucleon decay are sensitive to the particular decay modes. It is clear however that in general

$$\tau_N \geqq 10^{31} \text{ years} \quad ,$$

$$= .46 \times 10^{63} \text{ GeV}^{-1} \quad .$$

Assuming that the baryon number violating processes originate from a fundamental theory at a scale $M \gg m_W$ (i.e. much greater than the weak scale) and that the only particles with mass less than M are at m_W or below, we can, by integrating out all states with mass M or greater, obtain an effective action at m_W of the form

$$\mathcal{L}_{eff} \sim \frac{b}{M^{d-4}} \, O^{(d)} \quad .$$

The baryon violating operator of mass dimension d, $O^{(d)}$ is a product of local operators including particles with mass $\leqq O(m_W)$. $O^{(d)}$ is invariant under the low energy gauge group $SU_3 \times SU_2 \times U_1$. The nucleon lifetime is then given by

$$\tau_N = \frac{c^{-2}}{m_N}\left(\frac{M}{m_N}\right)^{2(d-4)}$$

where c is proportional to b and the matrix element of $O^{(d)}$ between the nucleon and particular final states.

For d = 6 and c ~ 1 we find that $M \geqq 4 \times 10^{15}$ Gev. Note that in the standard model with the minimal particle content, dimension 6 four fermi operators are the lowest dimension baryon violating operators consistent with $SU_3 \times SU_2 \times U_1$.[1] All such operators conserve the quantum number B-L (Baryon # - Lepton #). Dimension 6 operators require a new scale $M \sim 10^{15}$ Gev in order to give reasonable amounts of nucleon decay. In GUT's this scale is naturally associated with the grand unification scale, M_{GUT}.

Consider now d = 5 (4). If we assume $M \sim 4 \times 10^{15}$ Gev we find

$$c \leqq 10^{-16} \; (10^{-32})$$

in order to be consistent with observation.[F1]

It was realized that Baryon violating SUSY operators consistent with $SU_3 \times SU_2 \times U_1 \times SUSY$ can have d = 4,5 or 6.[2] Clearly for d = 4 or 5 either one must forbid these operators by introducing new symmetries or there must be some naturally small numbers inherent in c.

Dimension 4 SUSY operators will be eliminated by demanding a discrete symmetry R parity. d = 5 operators on the other hand can lead to reasonable nucleon decay rates.[3] All d = 5 operators (consistent with R-parity) conserve B-L.

The spectrum of states in the minimal low energy N = 1 SUSY theory is given in Table 1.

Table 1

ordinary particles		super partners	
spin 1/2		spin 0	
$q_i = \begin{pmatrix} u \\ d \end{pmatrix}_i^{1/3}$	$\bar{u}_i^{-4/3}$, $\bar{d}_i^{2/3}$	$\tilde{q}_i = \begin{pmatrix} \tilde{u} \\ \tilde{d} \end{pmatrix}_i^{1/3}$	$\tilde{\bar{u}}_i^{-4/3}$, $\tilde{\bar{d}}_i^{2/3}$
$l_i = \begin{pmatrix} \nu \\ e \end{pmatrix}_i^{-1}$	$\bar{e}_i^{+2}$	$\tilde{l}_i = \begin{pmatrix} \tilde{\nu} \\ \tilde{e} \end{pmatrix}_i^{-1}$	$\tilde{\bar{e}}_i^{+2}$
spin 1	$g, W^{\pm}, Z^{o}\gamma$	spin 1/2	$\tilde{g}, \tilde{w}^{\pm}, \tilde{\gamma}^{\alpha}$

spin 0

$$h = \begin{pmatrix} h^+ \\ h^0 \end{pmatrix}^{+1} \qquad \bar{h} = \begin{pmatrix} \bar{h}^- \\ \bar{h}^0 \end{pmatrix}^{-1}$$

$Q = I_3 + Y/2$ (hypercharge is given in superscripts)
$i = 1,2,3$ number of generations
$\alpha = 1,\ldots, 4$ number of neutralinos
$\tilde{w}^{\pm}$, $\tilde{\gamma}^{\alpha}$ are superpositions of gaugino and higgsino states.

It will be convenient to use the compact superspace notation to describe the baryon violating operators. We thus define a general chiral superfield

$$\Phi(x_\mu,\theta_\alpha) = \phi(x) + \sqrt{2}(\theta\psi(x)) + (\theta\theta)\, F(x)$$

where $\phi(x)$, $\psi(x)$ are a complex scalar and left-handed fermi field, respectively, and $F(x)$ is a complex auxiliary field. θ_α is a two-component left-handed Weyl spinor. Note Φ transforms as a Lorentz scalar and satisfies commutation relations. For gauge fields we introduce the chiral SUSY field strength (in the Wess-Zumino gauge)

$$W_\alpha(x_\mu,\theta_\beta) = -i\lambda_\alpha(x) + [D(x) - i\,\sigma^{\mu\nu}F_{\mu\nu}(x)]\theta_\alpha + (\theta\theta)\,\sigma^\mu\,\partial_\mu\,\bar{\lambda}(x)$$

where $\lambda_\alpha(x)$, $F_{\mu\nu}(x) \equiv \partial_\mu A_\nu - \partial_\nu A_\mu + [A_\mu, A_\nu]$ are the gaugino and gauge field strength, respectively and $D(x)$ is an auxiliary field. The superfield content of the minimal low energy theory is given in Table II.

Table II

quarks, leptons
squarks, sleptons

$$Q_i = \begin{pmatrix} U \\ D \end{pmatrix}_i \qquad \bar{U}_i, \quad \bar{D}_i$$

$$L_i = \begin{pmatrix} V \\ E \end{pmatrix}_i \qquad \bar{E}_i$$

gauge bosons, gauginos

W^3_α, W^2_α, W^1_α for SU_3, SU_2, U_1 resp.

Higgs bosons, Higgsinos

$$H = \begin{pmatrix} H^+ \\ H^o \end{pmatrix} \qquad \bar{H} = \begin{pmatrix} \bar{H}^- \\ \bar{H}^o \end{pmatrix}$$

The baryon number violating operators consistent with $SU_3 \times SU_2 \times U_1 \times$ SUSY are given in Table III.

Table III

Baryon and/or Lepton number violating operators

	$\underline{d = 6}$	
R_{even}	$\frac{b_1}{M^2} \int d^4\theta \; Q^*_{\alpha i} \bar{U}_j \; Q^*_{\beta k} \; \bar{E}_\ell \; \epsilon^{\alpha\beta}$	$\Delta B = \Delta L$
R_{even}	$\frac{b_2}{M^2} \int d^4\theta \; Q^*_{\alpha i} \bar{U}_j \; L^*_{\beta k} \; \bar{D}_\ell \; \epsilon^{\alpha\beta}$	$\Delta B = \Delta L$
	$\underline{d = 5}$	
R_{even}	$\frac{b_3}{M} \int d^2\theta \; Q_i Q_j Q_k L_\ell$	$\Delta B = \Delta L$
R_{even}	$\frac{b_4}{M} \int d^2\theta \; \bar{U}_i \bar{U}_j \bar{D}_k \bar{E}_\ell$	$\Delta B = \Delta L$
	$\int d^2\theta \; Q_i Q_j Q_k \bar{H}$	$\Delta B = 1 \quad \Delta L = 0$
	$\int d^4\theta \; Q_i Q_j \bar{D}^*_k$	$\Delta B = 1 \quad \Delta L = 0$
	$\int d^4\theta \; Q_i \bar{U}_j L^*_k$	$\Delta B = 0 \quad \Delta L = -1$
	$\underline{d = 4}$	
	$\int d^2\theta \; \bar{U}_i \bar{D}_j \bar{D}_k$	$\Delta B = -1 \; \Delta L = 0$
	$\int d^2\theta \; Q_i \bar{D}_j L_k$	$\Delta L = 1 \quad \Delta B = 0$
	$\int d^2\theta \; \bar{E}_i L_j L_k$	$\Delta L = 1 \quad \Delta B = 0$

In order to avoid catastrophic nucleon decay rates we shall require our theory to have the discrete symmetry

$$\Phi(x,\theta) \rightarrow \eta_\Phi \, \Phi(x,\theta)$$

$$W_\alpha(x,\theta) \rightarrow W_\alpha(x,\theta)$$

where

$$\eta_\Phi = -1 \text{ for } \Phi = Q, \bar{U}, \bar{D}, L, \bar{E}; \; \eta_\Phi = +1 \text{ otherwise} \quad .$$

This is the so-called family reflection symmetry.[4] It is equivalent to the discrete symmetry

$$(-1)^{3(B-L)} \quad .$$

Note that the discrete symmetry, R-parity,[5] given by

$$\Phi(x,\theta) \rightarrow \eta_\Phi \, \Phi(x,-\theta)$$

$$W_\alpha(x,\theta) \rightarrow - W_\alpha(x,-\theta)$$

with η_Φ as above is equivalent to the discrete symmetry

$$(-1)^{3(B-L)} \cdot (-1)^F$$

where F is fermion number. Under R-parity all ordinary particles are R even and all superpartners are R odd.

Since $(-1)^F$ is conserved in any Lorentz invariant theory we see that R-parity and the Family Reflection symmetry are identical. Imposing this symmetry eliminates all the unwanted d = 4 operators. It also eliminates some d = 5 operators. All remaining operators satisfy, as previously stated, $\Delta B = \Delta L$.

The remaining dimension 5 operators

$$F_{ijkl} = \int d^2\theta \; Q_{\alpha ai} \; Q_{\beta bj} \; Q_{\gamma ck} \; L_{\delta 1} \times \epsilon^{\alpha\beta} \; \epsilon^{\gamma\delta} \; \epsilon^{abc}$$

$$\bar{F}_{ijkl} = \int d^2\theta \; \bar{U}_i^a \; \bar{U}_j^b \; \bar{D}_k^c \; \bar{E}_1 \; \epsilon_{abc}$$

where $(\alpha\beta\gamma\delta)$ (a,b,c) and (i,j,k,l) are SU_2, SU_3 and generation indices resp., satisfy

$$F_{iiji} \equiv 0 \quad , \quad \bar{F}_{ijii} \equiv 0 \text{ for } i = j.$$

Therefore the dominant nucleon decay modes are expected to have strange particles in the final state.[3]

Note, when SUSY is broken at a scale of order m_W then dimension 6 SUSY operators with SUSY breaking insertions can lead to lower dimensional operators in terms of component fields. For example with the insertion of the SUSY breaking term $m^2\theta^2\bar{\theta}^2$, responsible for scalar masses, into the dimension 6 operators we obtain a dimension four (four scalar) operator of the form

$$\frac{bm^2}{M_{GUT}^2} \; \phi^2\phi^{*2} \quad .$$

Such terms contribute negligible nucleon decay rates since they require at least two loops worth of dressing in order to obtain an effective four fermi operator.[6]

Minimal SU_5 SUSY GUT

The spectrum of states in the minimal SU_5 SUSY GUT is given in Table IV.

Table IV

$$10_i \supset \begin{pmatrix} \bar{U} & Q \\ & \bar{E} \end{pmatrix}_i \qquad \bar{5}_i \supset \begin{pmatrix} \bar{D} \\ L \end{pmatrix}_i$$

$$W^5_\alpha$$

$$H_5 = \begin{pmatrix} H_3 \\ H \end{pmatrix} \qquad \bar{H}_5 = \begin{pmatrix} \bar{H}_3 \\ \bar{H} \end{pmatrix}$$

$$24$$

SU_5 breaking

The 24 is necessary to break SU_5 to $SU_3 \times SU_2 \times U_1$ at M_{GUT}. It is also used to give mass to the color triplet Higgs H_3, $\bar{H}_3$ at a scale $\sim 0\,(M_{GUT})$. The Higgs doublets H and $\bar{H}$ obtain mass $\mu \sim 0(m_w)$. This is accomplished by fine tuning parameters. In other scenarios this splitting may be accomplished more naturally.[7] The low energy theory $E \ll M_{GUT}$ includes only those states given in Tables I and II.

The standard renormalization group analysis from M_{GUT} to m_W determines the parameters M_{GUT} and $\alpha_5(M_{GUT})$ in terms of $\alpha_3(m_W)$ and $\alpha_{EM}(m_W)$ and predicts the value of $\sin^2\hat{\theta}_W(m_W)$. We find (at two loops, assuming a common SUSY threshold at M_W)[8]

$$M_{GUT} \cong 6 \times 10^{16}\left(\frac{\Lambda_{\overline{ms}}}{1\text{Gev}}\right) \text{Gev}$$

$$\alpha_5(M_{GUT}) \sim 1/25$$

and

$$\sin^2\hat{\theta}_W(m_W) = .233 \begin{matrix} + .006 \\ - .001 \end{matrix}$$

with the upper (lower) error given by $\Lambda_{\overline{ms}} = .1(.4)$ Gev and the central value given by $\Lambda_{\overline{ms}} = .3$ Gev. Note that M_{GUT} in SUSY SU_5 is about 40 times larger than M_{GUT} in ordinary SU_5, and $\sin^2\hat{\theta}_W$ is about 10% larger than the ordinary SU_5 prediction $\sin^2\hat{\theta}_W(m_W) = .209 \pm {}^{.007}_{.001}$. It is worth re-remarking that the experimental value of $\sin^2\hat{\theta}_W(m_W)$ appears to be increasing. Recent data from CDHSW, CHARM, CCFRR and FMM on $\nu_\mu N$ deep inelastic scattering give $\sin^2\hat{\theta}_W(m_W) = .232 \pm .004 \pm .006$ and from UA2 using the W and Z masses $\sin^2\hat{\theta}_W(m_W) = .232 \pm .004 \pm .008$. These results have significantly smaller errors than previous experiments.[9]

Nucleon decay

Dimension 6 operators

There are only two such operators which are given in Table III. They can result from either gauge boson exchange or scalar Higgs triplet exchange. (see Fig. 1.)

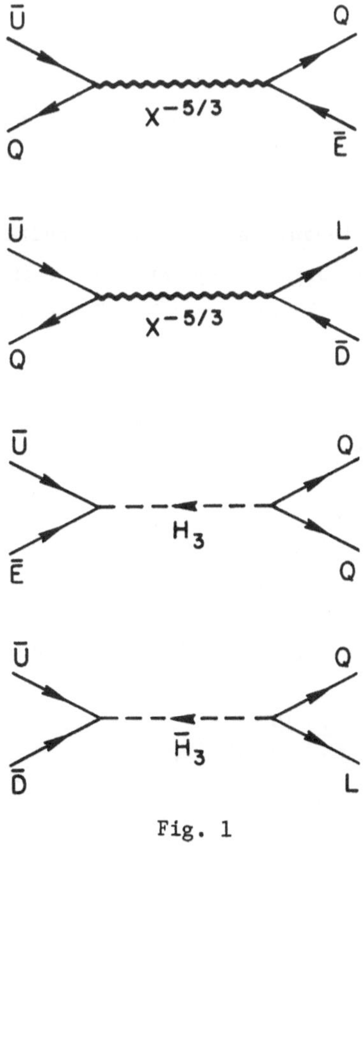

Fig. 1

Gauge exchange

Since M_{GUT} has increased in SUSY SU_5 we find that τ_N due to gauge exchange alone is unobservably long.

$$\tau_N = 2 \left(\frac{\bar{\Lambda}_{ms}}{.1 \text{ Gev}}\right)^5 10^{34\pm1} \text{years}$$

Hence the standard dominant decay mode $p \to \pi^o e^+$ is suppressed.

Higgs exchange

Higgs scalar exchange can be important for Higgs masses of order 10^{10-11} Gev. These processes are naturally suppressed due to small Yukawa couplings. Although it is possible to have such light Higgs scalars, naturally, in a SUSY SU_5 model (see Geometric Hierarchy),[10] they are not light in the minimal version of SUSY SU_5. They have mass of order M_{GUT}. The dominant decay modes for Higgs scalar exchange are

$$P \to K^o \mu^+ \ , \ K^+ \bar{\nu}_\mu$$

$$n \to K^o \bar{\nu}_\mu \quad .$$

Dimension 5 operators

There are two relevant dimension 5 operators in an R-parity invariant low energy theory. They are given in Table III. They are obtained by exchanging the color triplet Higgs fermions H_3 and $\bar{H}_3$ (see Fig. 2), using the following Yukawa interactions.

$$L_{Yukawa} \supset \int d^2\theta \left(H_3 \{ D\left(V^T \frac{m_u}{v}\right) U + U\left(\frac{m_u}{v} V\right) D \right.$$

$$+ \bar{E}\left(V^T \frac{m_u}{v}\right) \bar{U} + \bar{U}\left(\frac{m_u}{v} V\right) \bar{E} \}$$

$$\left. + \bar{H}_3 \{ -V\left(\frac{m_d}{\bar{v}}\right) D + E\left(\frac{m_d}{\bar{v}} V^\dagger\right) U + \bar{D}\left(\frac{m_d}{\bar{v}} V^\dagger\right) \bar{U} \} \right)$$

where V_{ij} is the Cabibbo-Kobayashi-Maskawa matrix, $(\bar{m}_u, \bar{m}_d)$ are diagonal mass matrices and $(v, \bar{v})$ given by $v = \langle H_0 \rangle$, $\bar{v} = \langle \bar{H}_0 \rangle$ are weak vacuum expectation values.

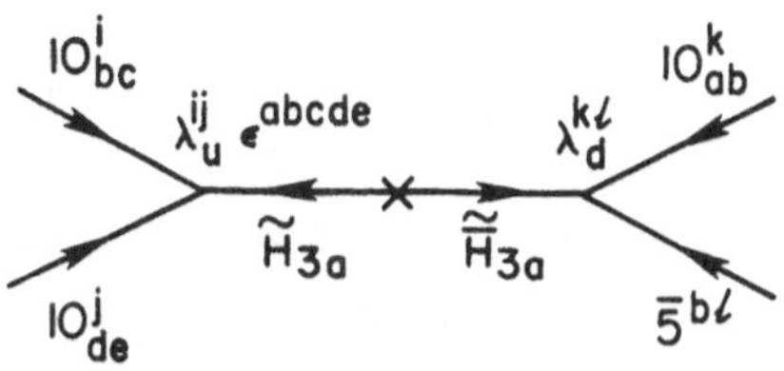

Fig. 2

We thus obtain the effective dimension 5 operators

$$\mathcal{L}_{\Delta B} \sim \frac{\Delta m_{3\bar{3}}}{M^2} \left\{ \left[D\left(V^T \frac{m_u}{v}\right) U + U\left(\frac{m_u}{v} V\right) D \right] \times \left[V\left(\frac{m_d}{\bar{v}}\right) D + E\left(\frac{m_d}{\bar{v}} V^\dagger\right) U \right] \right.$$

$$\left. + \left[\bar{E}\left(V^T \frac{m_u}{v}\right) \bar{U} + \bar{U}\left(\frac{m_u}{v} V\right) \bar{E} \right] \times \left[\bar{D}\left(\frac{m_d}{\bar{v}} V^\dagger\right) \bar{U} \right] \right\}_{F\ \text{term}}$$

where $\Delta m_{3\bar{3}}$ is the $H_3 - \bar{H}_3$ mixing mass term and M is the average $H_3 - \bar{H}_3$ mass. In the minimal SUSY SU_5 model $M \equiv \Delta m_{3\bar{3}}$. Since H_3 and $\bar{H}_3$ are the SU_5 partners of the Higgs doublets H and $\bar{H}$ the couplings are known and are small, i.e. the couplings are proportional to the quark masses. This is yet another reason for the dominance of strange particle final states.

The dimension 5 operators are quadratic in scalar and in fermi fields. They are dressed at the weak scale by gaugino exchange in order to obtain effective four fermi operators contributing to nucleon decay.

The low energy theory; $E << M_{GUT}$

SUSY Breaking

The gaugino, Higgsino, squark and slepton masses and mixing angles determine the parameters in the effective 4-fermi baryon number violating operators. A brief review of the low energy SUSY mass spectrum will suffice for our purposes.[11]

Gauginos-Higgsinos

Assuming a common SUSY breaking mass term for gauginos at a scale $\mu \sim 2 \times 10^{18}$ Gev we find the SU_3, SU_2, U_1 gaugino majorana mass terms at m_W in the ratio $M_3:M_2:M_1 = \alpha_3:\alpha_2:\frac{5}{3}\alpha_1$.

The gluino mass is M_3. The charged and neutral winos and Higgsinos mix at m_W. The mass matrices are given in Table V.

Table V

$$\begin{array}{c|cc} & \tilde{w}^+ & \tilde{h}^+ \\ \hline \tilde{w}^- & M_2 & g_2 v \\ \tilde{\bar{h}}^- & g_2\bar{v} & \epsilon \end{array}$$

$$\begin{array}{c|cccc} & \tilde{w}^3 & \tilde{b} & \tilde{h}^o & \tilde{\bar{h}}^o \\ \hline \tilde{w}^3 & M_2 & 0 & g_2 v & -g_2\bar{v} \\ \tilde{b} & 0 & M_1 & g_1 v & g_1\bar{v} \\ \tilde{h}^o & g_2 v & g_1 v & 0 & \epsilon \\ \tilde{\bar{h}}^o & -g_2\bar{v} & g_1\bar{v} & \epsilon & 0 \end{array}$$

$$m_W^2 \equiv g_2^2 \frac{(v^2+\bar{v}^2)}{2}, \quad m_Z^2 = m_W^2/\cos^2\theta_W$$

In the limit $M_2 \sim \epsilon < m_W$ the charged states have predominantly a Dirac mass of order m_W. The neutralinos have approximate mass eigenstates given by a photino $\tilde{\gamma}$, with mass $m_{\tilde{\gamma}} \sim \frac{8}{3}\frac{\alpha}{\alpha_3} m_{\tilde{g}}$ or $m_{\tilde{\gamma}} \sim \frac{1}{5} m_{\tilde{g}}$; a Higgsino with mass $\sim\epsilon$ and a Dirac Zino with mass $\sim m_Z$. The experimental limit on gluino masses from UA1 is now $m_{\tilde{g}} \geqq 60$ Gev. Thus $M_2 \geqq 20$ Gev which is still compatible with the above limit.

Squarks and Sleptons

Squark and slepton masses come from several different sources. For definiteness let us focus on the top squark mass matrix. It is given in Table VI.

Table VI

$$\begin{array}{ccc} & \tilde{t} & \tilde{\bar{t}}^{*} \\ \tilde{t}^{*} & m_{\tilde{t}}^2 + \frac{\left(g_2^2 - \frac{1}{3}g_1^2\right)}{4}\left(v^2 - \bar{v}^2\right) + m_t^2 & Am_t \\ \tilde{\bar{t}} & Am_t & m_{\tilde{\bar{t}}}^2 + \frac{g_1{}^2}{3}\left(v^2 - \bar{v}^2\right) + m_t{}^2 \end{array}$$

There are four different contributions. $m_{\tilde{t}}$ and $m_{\tilde{\bar{t}}}$ are SUSY breaking terms which are assumed to be equal at M. At m_W this contribution is ~ 3 times larger for squarks than for sleptons due to radiative corrections coming from gluinos. For the top squark $m_{\tilde{t}}$ and $m_{\tilde{\bar{t}}}$ may also receive large corrections due to a large Yukawa coupling λ_t. The A parameter is a soft SUSY breaking term for cubic scalar interactions. The terms proportional to $(v^2-\bar{v}^2)$ come from SUSY gauge interactions --so-called D terms. This contribution differs for up and down squarks. Finally m_t is the supersymmetric contribution.

Nucleon decay

The dominant four fermi operators result from the processes depicted in Figs. 3, 4 and 5.

In order to make contact with experiment we must take into account two additional effects. We must find the renormalization group corrections to these effective four fermi operators from the physics between M_{GUT} and m_N (the nucleon mass). This is made in two stages, a) from M_{GUT} to m_W for which we may renormalize the effective SUSY dimension 5 operators and b) from m_w to m_N which is identical to the corrections occuring in non-SUSY theories.[12] Note that we are assuming in this analysis that the SUSY threshold occurs at m_W. Secondly, we must take matrix elements of the effective four fermi operators between the nucleon and appropriate final states. This requires knowledge of strong QCD effects which introduces large theoretical uncertainties. We shall rely on the chiral Lagrangian approach to obtain an estimate of the amplitudes[13] (see Fig. 6). This introduces the strong interaction parameter β describing the three quark fusion process of (Fig. 6b). Theoretical estimates of β are in the range $.003 < \beta < .03\ \mathrm{GeV}^3$.

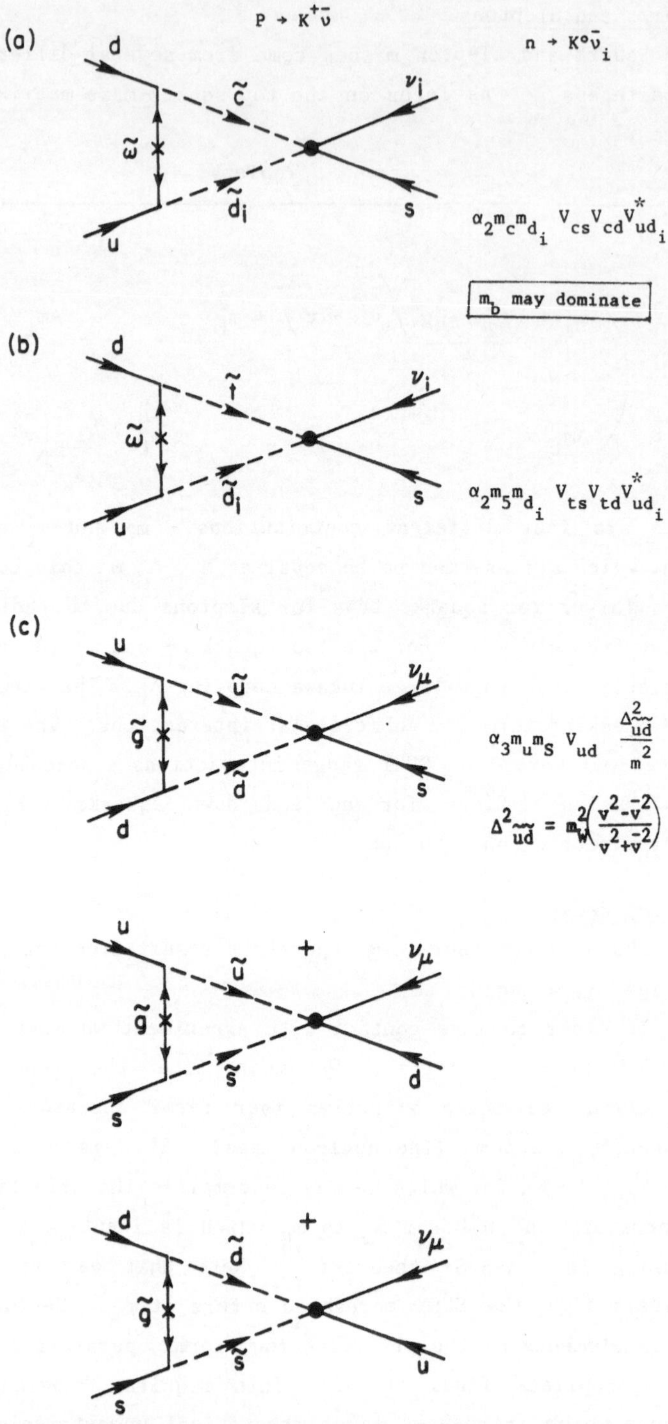
P → K⁺ν̄
n → K°ν̄ᵢ
(a)
d
c̃
νᵢ
ω̃
d̃ᵢ
u
s
α₂m_c m_dᵢ V_cs V_cd V*_udᵢ
m_b may dominate
(b)
d
t̃
νᵢ
ω̃
d̃ᵢ
u
s
α₂m₅m_dᵢ V_ts V_td V*_udᵢ
(c)
u
ũ
ν_μ
g̃
d̃
d
s
α₃m_u m_S V_ud Δ²_ũd̃ / m²
Δ²_ũd̃ = m²_W((v² − v̄²)/(v² + v̄²))
u
+
ũ
ν_μ
g̃
s̃
s
d
d
+
d̃
ν_μ
g̃
s̃
s
u

Fig. 3

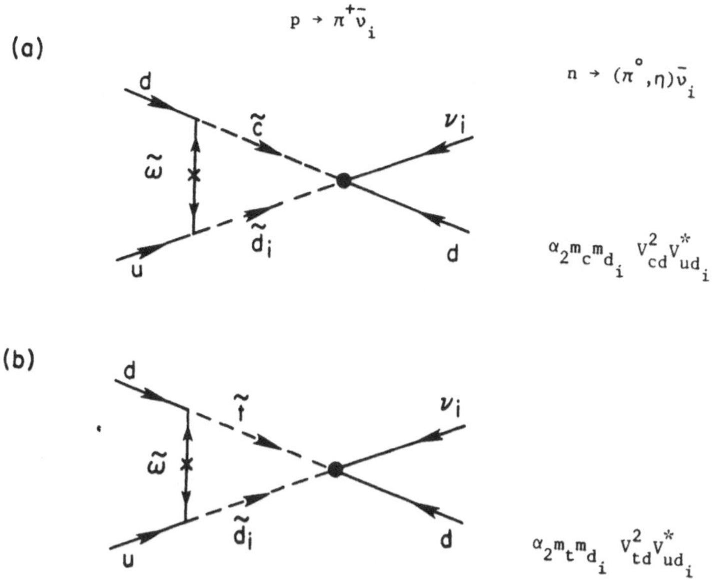

$p \to \pi^+\bar{\nu}_i$
(a)
$n \to (\pi^\circ, \eta)\bar{\nu}_i$
d
$\tilde{c}$
ν_i
$\tilde{\omega}$
$\tilde{d}_i$
u
d
$\alpha_2 m_c m_{d_i}\ V^2_{cd} V^*_{ud_i}$
(b)
d
$\tilde{t}$
ν_i
$\tilde{\omega}$
$\tilde{d}_i$
u
d
$\alpha_2 m_t m_{d_i}\ V^2_{td} V^*_{ud_i}$

Fig. 4

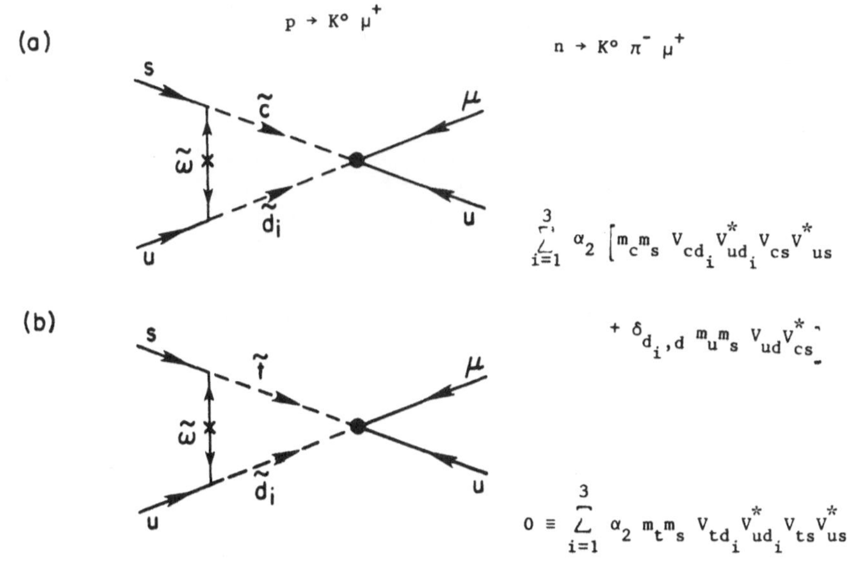

$p \to K^\circ \mu^+$
(a)
$n \to K^\circ \pi^- \mu^+$
s
$\tilde{c}$
μ
$\tilde{\omega}$
$\tilde{d}_i$
u
u
$\sum_{i=1}^{3} \alpha_2 \left[m_c m_s\ V_{cd_i} V^*_{ud_i} V_{cs} V^*_{us} \right.$
(b)
$\left. + \delta_{d_i,d}\ m_u m_s\ V_{ud} V^*_{cs} \right]$
s
$\tilde{t}$
μ
$\tilde{\omega}$
$\tilde{d}_i$
u
u
$0 \equiv \sum_{i=1}^{3} \alpha_2\ m_t m_s\ V_{td_i} V^*_{ud_i} V_{ts} V^*_{us}$

Fig. 5

a

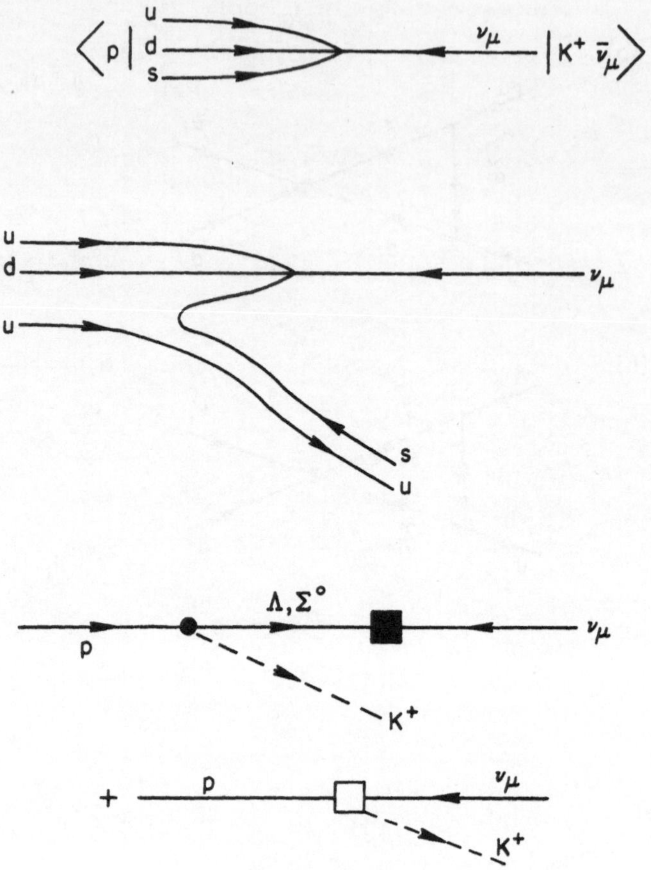

b $\beta \equiv \langle \Sigma^{\circ} \,|\, uds \,|\, 0 \rangle$

Fig. 6

Table VII

Branching ratios for some dominant nucleon decay modes

Chadha and Daniel[14]

$p \to K^+(\bar{\nu}_\mu, \bar{\nu}_\tau)$	90%
$p \to \pi^+(\bar{\nu}_\mu, \bar{\nu}_\tau)$	10%
$p \to K^0\mu^+$	$\sim .6\times10^{-3}$
$p \to \pi^0\mu^+$	$\sim .3\times10^{-3}$
$n \to K^0(\bar{\nu}_\mu, \bar{\nu}_\tau)$	95%
$n \to \pi^0(\bar{\nu}_\mu, \bar{\nu}_\tau)$	3%
$n \to \eta(\bar{\nu}_\mu, \bar{\nu}_\tau)$	2%
$n \to K^0\bar{\nu}_e$	$\sim .3\times10^{-3}$

In Table VII we present branching ratios for nucleon decay due to the processes given in Figs. 3,4,5, assuming $\tilde{w}$ dominance. Several comments are in order.

A) The sum of graphs in Fig. 3c has been shown to vanish for degenerate squarks.[15] However in supergravity theories the up-down squark mass difference is given by $\Delta^2_{\tilde{u}\tilde{d}} = m_W^2\left(\frac{v^2-\bar{v}^2}{v^2+\bar{v}^2}\right)$ which can be of order the squark masses.

As a result the ratio of amplitudes

$\frac{A_{\tilde{w}}(p\to K^+\bar{\nu})}{A_{\tilde{g}}(p\to K^+\bar{\nu})}$ can be of order one.[16]

B) It has been shown by Arnowitt, Chamseddine and Nath[17] that for a range of parameters with, in particular $m_t \geqq 40$ Gev that the amplitudes of Figs. 3a and 3b can be arranged to cancel. If one then also suppresses the gluino exchange diagrams of Fig. 3c by increasing the average squark mass one finds the dominant nucleon decay modes are

$$n \to \pi_0 \bar{\nu}$$

$$p \to \pi^+ \bar{\nu} \ .$$

C) I have ignored neutralino exchanges since these are typically negligible except in the case that the aforementioned cancellations occur. In this case neutralino exchange can be suppressed with

$$\frac{v^2 - \bar{v}^2}{v^2 + \bar{v}^2} \sim 1.$$

D) Finally, for large top quark masses, the graphs of Fig. 7 become important. Large top masses lead to large $\tilde{t} - \tilde{\bar{t}}$ mixing for $A \sim m$ as needed in Fig. 7b or to large $\tilde{t} - \tilde{u}$ mixing due to large renormalization group corrections to $m^2_{\tilde{t}}$ and $m^2_{\tilde{\bar{t}}}$ which are proportional to the top quark Yukawa coupling λ_t. The ratio of rates from Fig. 7(A and C) to Fig. 4a is of order[18]

$$\frac{\Gamma_{gluino}(p \to K^0 \mu^+)}{\Gamma_{\tilde{w}}(p \to K^+ \bar{\nu})} \cong \frac{1}{10} \left(\frac{\alpha_3 m_t \ V_{ts} \ V_{td}}{\alpha_2 m_c \ V^2_{cd}} \left(\frac{\Delta^2_{\tilde{u}\tilde{t}}}{m^2} \right) \frac{m_{\tilde{g}}}{m_{\tilde{w}}} \right)^2 .$$

Assuming $\frac{V_{ts} \ V_{td}}{V^2_{cd}} \sim \sin^4\theta_c$, and using $m_{\tilde{g}}/m_{\tilde{w}} \sim \alpha_3/\alpha_2$ we obtain $\Delta^2_{\tilde{u}\tilde{t}}/m^2 \sim 1$ and

$$\frac{\Gamma_{gluino}(p \to K^0 \mu^+)}{\Gamma_{wino}(p \to K^+ \bar{\nu})} \cong 1$$

for $m_t = 40$ Gev.

Note Fig. 7c illustrates that the right handed operator of the form $\bar{U}_i \bar{U}_j \bar{D}_k \bar{\mu}_\ell$ can contribute to nucleon decay in this limit.

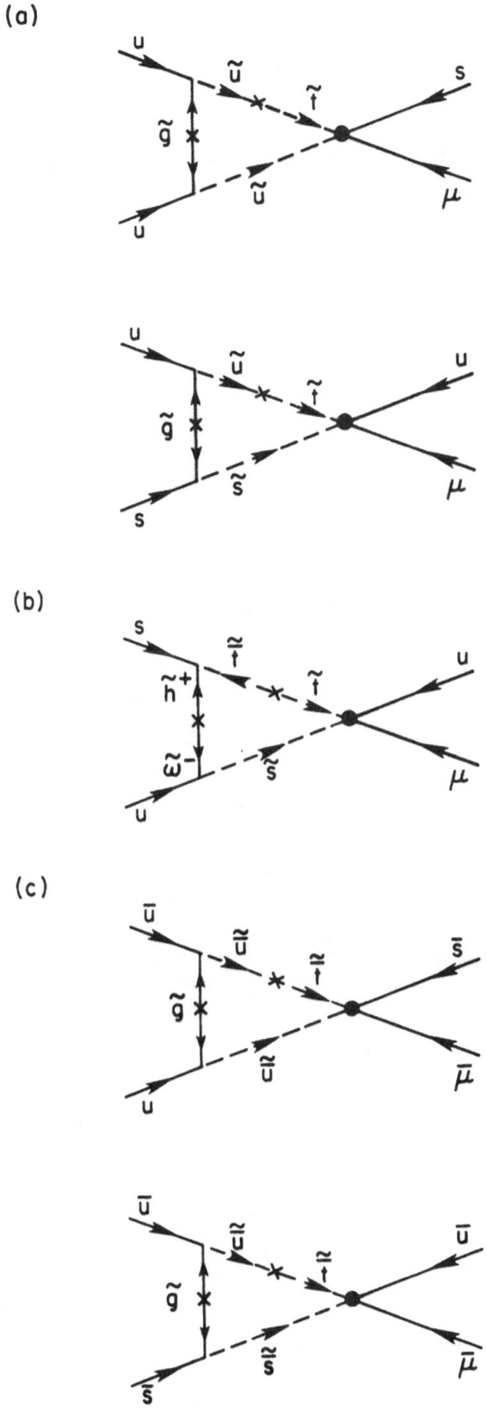

Fig. 7

The dominant decay mode of the proton is $P\to K^{+}\bar{\nu}_{\mu}$. The decay rate for this mode (assuming $\tilde{w}$ dominance) is given by [17]

$$\Gamma(p\to K^{+}\bar{\nu}_{\mu}) \geq \frac{\beta^2}{M^2}\frac{m_N}{32\pi f_\pi^2}\left(1-\frac{m_K^2}{m_N^2}\right)^2 A_L^2\left(A_S^L\right)^2 \left| 1 + \left(\frac{m_N}{m_\Lambda}\right)(D+F)\right|^2 \left|A_{K\bar{\nu}_\mu}\right|^2$$

where

$$A_{K\bar{\nu}_\mu} = \frac{\alpha_2^2\, m_c\, m_s \sin^2\Theta_c}{2m_w^2 \sin 2\Theta_h}\left[F(\tilde{c};\tilde{s};\tilde{w}) + F(\tilde{c};\tilde{\mu};\tilde{w})\right]$$

and

$$\tan\Theta_h \equiv \frac{\langle h^0\rangle}{|\langle \bar{h}^0\rangle|} .$$

D, F and f_π are parameters in the chiral Lagrangian. We take

$D = .76$

$F = .48$

$f_\pi = 139$ Mev.

Finally, the function F (a result of the loop integration) is given by

$$F(m_1, m_2, \tilde{w}) \equiv \cos^2\Theta\, f(m_1, m_2, m_+) + \sin^2\Theta\, f(m_1, m_2, m_-)$$

where

$$\tan\Theta = \frac{(M_2+\epsilon) - \sqrt{(M_2+\epsilon)^2 + 4m_w^2}}{2m_W}$$

$$m_\pm = \frac{1}{2}\left[M_2-\epsilon \pm \sqrt{(M_2+\epsilon)^2 + 4m_w^2}\right]$$

and

$$f(m_1, m_2, m_3) = \frac{m_3}{m_2^2-m_3^2}\left[\frac{m_2^2}{m_1^2-m_2^2}\,\ell n\,\frac{m_1^2}{m_2^2} - \frac{m_3^2}{m_1^2-m_3^2}\,\ell n\,\frac{m_1^2}{m_3^2}\right] .$$

Note this is not the most general result for the masses $m_\pm$ and mixing angles but is valid in the limit $\tan\Theta_h = 1$. $m_\pm$ are the charged wino eigenstates (see Table V) and Θ is the appropriate mixing angle. The result for $\Gamma(p\to K^{+}\bar{\nu}_{\mu})$ is clearly dependent on the low energy supersymmetric mass spectrum and the mass M of the heavy color triplet Higgs fermions. If we insert some characteristic values for the experimental limit for $\tau_{p\to K^{+}\bar{\nu}_{\mu}}$ we can obtain a lower bound for M.

We find

$$M \cong 10^{17\pm 1}\left(\frac{M_2}{20 \text{ Gev}}\right)\left(\frac{100 \text{ Gev}}{m_{\tilde{q}}}\right)^2 \text{ Gev}$$

for $m_{\tilde{q}} \sim m_{\tilde{\ell}} >> M_2 \sim \epsilon$.

Since we expect $M \sim M_{GUT}$ and $M_{GUT} \sim 10^{16}$ Gev in the minimal low energy supersymmetric model, we conclude that nucleon decay should be observable in SUSY GUTS.

In fact, once the low energy supersymmetric spectrum is observed, then it may be possible to rule out the minimal SU_5 SUSY GUT on the basis of nucleon decay. There is then only one free parameter in the theory (the color triplet Higgs mass M) and this parameter cannot be much greater than M_{GUT}, as remarked previously.[19]

$\underline{M_{GUT} \sim M_{pl}}$.

It might be preferable to have a theory for which $M_{GUT} \sim M_{pl}$. This appears to be the case in superstring theories. Moreover it has the aesthetic advantage of containing the ultimate desert with no new physics between m_W and M_{pl}. It may also be forced upon us by experiment. It is worth pointing out that in order to have $M_{GUT} \sim M_{pl}$ one must necessarily have additional light states.

A particularly interesting possibility occurs in models where SUSY breaking and the GUT symmetry breaking have a common origin. In this case one finds the additional states in Table VIII.[20]

Table VIII

	$SU_3 \times SU_2 \times U_1$
$\underset{\sim}{8}$	(8, 1, 0)
$\underset{\sim}{3}$	(1, 3, 0)
$\bar{E}$	(1, 1, 2)
E	(1, 1, -2)

Using the standard renormalization group analysis one finds[20]

$$M_{GUT} \sim 10^{19\pm 2} \text{ Gev}$$

$$\sin^2\hat{\theta}_w \sim .230 \pm .005$$

where the errors are theoretical and include the standard threshold uncertainties.

Conclusion

Nucleon decay in SUSY GUT's has a characteristic signature. One expects

$$p \to K^{+}\bar{\nu}_{\mu}$$

$$n \to K^{0}\bar{\nu}_{\mu}$$

to be the dominant decay modes. Unlike the case of ordinary SU_5 GUT's however, the predictions in SUSY theories depend on the details of the as yet unobserved "low energy" supersymmetric spectrum and also on the mass M of color triplet Higgs fermions with $M \sim M_{GUT}$. There are ranges in parameter space for which the decay modes

$$P \to \pi^{+}\bar{\nu}_{\mu}$$

$$\to K^{0}\mu^{+}$$

$$n \to \pi^{0}\bar{\nu}_{\mu}$$

$$\to \eta\bar{\nu}_{\mu}$$

may be significant or may even dominate. Assuming the "low energy" SUSY spectrum is fixed experimentally one may then confidently calculate (in a minimal SU_5 SUSY GUT) the nucleon decay rate as a function of M. This free parameter has an upper bound obtained by requiring the theory to be perturbative at the scale M_{GUT}. Hence an absolute theoretical upper bound for τ_N may in principle be obtained and compared with experiment.

We remind the reader that in order to obtain models for which M_{GUT} is of order M_{pl}, as may be desired for example in superstring theories, one necessarily expects additional "low energy" states (see Table VIII) beyond the minimal low energy SUSY spectrum (see Tables I & II).

Finally we note that minimal SUSY GUT's typically predict a value for $\sin^2\hat{\theta}_w(m_w)$ of order .233. This is significantly higher than minimal non-SUSY GUT's. It is a nice test for the simple models discussed here.

References

1. J. Pati and A. Salam, Phys. Rev. D10, 275 (1974); H. Georgi and S. L. Glashow, Phys. Rev. Lett. 32, 438 (1974); H. Georgi, H. Quinn and S. Weinberg, Phys. Rev. Lett. 33, 451 (1974); A. J. Buras et. al., Nucl. Phys. B135, 66 (1978); S. Weinberg, Phys. Rev. Lett. 43, 1566 (1979); F. Wilczek and A. Zee, Phys. Rev. Lett. 43, 1571 (1979). See also the review by P. Langacker, Phys. Rep. 72C, 185 (1981).

2. N. Sakai and T. Yanagida, Nucl. Phys. B197, 523 (1982); S. Weinberg, Phys. Rev. D26, 257 (1982).

3. S. Dimopoulos, S. Raby and F. Wilczek, Phys. Lett. 112B, 133 (1982).

4. S. Dimopoulos and H. Georgi, Nucl. Phys. B193, 150 (1981); N. Sakai, Z. f. Phys. C11, 153 (1981).

5. P. Fayet, Nucl. Phys. B90 104 (1975); L. O'Raifeartaigh, Lecture Notes on Supersymmetry, Comm. Dias. Series A, no. 22; G. Farrar and S. Weinberg, Phys. Rev. D27, 2732 (1983).

6. J.-P. Derendinger and C. A. Savoy, Phys. Lett. 118B, 347 (1982); N. Sakai, Phys. Lett. 121B, 130 (1983).

7. S. Dimopoulos and F. Wilczek, Proc. of the 1981 Erice Summer School on Particle Physics, A. Zichichi, ed.; B. Grinstein, Nucl. Phys. B206, 387 (1982); A. Masiero, D. V. Nanopoulos, K. Tamvakis, and T. Yanagida, Phys. Lett. 115B, 380 (1982).

8. S. Dimopoulos, S. Raby, and F. Wilczek, Phys. Rev. D24, 1681 (1981); S. Dimopoulos and H. Georgi, ref. 4; M. B. Einhorn and D. R. T. Jones, Nucl. Phys. B196, 475 (1982); W. J. Marciano and G. Senjanovic, Phys. Rev. D25, 3420 (1982); L. E. Ibañez and G. G. Ross, Phys. Lett. 105B, 439 (1981).

9. For a review, see the talk by G. Altarelli at the Berkeley conference, July 22, 1986.

10. S. Dimopoulos and S. Raby, Nucl. Phys. B219, 479 (1983); S. Raby, Proc. of the Winter School at Mahabaleshwar, India, Jan. 1984, Lecture Notes in Physics, Springer-Verlag, Berlin (1984).

11. For a review, see H. Haber and G. Kane, Phys. Rep., 117, 75 (1985); H. P. Nilles, Phys. Rep. 110 (1985).

12. J. Ellis, D. V. Nanopoulos and S. Rudaz, Nucl. Phys. B202 43 (1982).

13. M. Claudson, M. B. Wise and L. J. Hall, Nucl. Phys. B195 297 (1982).

14. S. Chadha and M. Daniel, Phys. Lett. 137B 374 (1984).

15. V. M. Belyaev and M. I. Vysotsky, Phys. Lett. 127B 215 (1983).

16. J. Milutinovic, P. B. Pal and G. Senjanovic, Phys. Lett. 140B, 324 (1984).

17. P. Nath, A. H. Chamseddine and R. Arnowitt, Phys. Rev. D32 2348 (1985).

18. S. Chadha, G. D. Coughlan, M. Daniel and G. G. Ross, Phys. Lett. 149B 477 (1984).

19. K. Enqvist, A. Masiero and D. V. Nanopoulos, Phys. Lett. 156B, 209 (1985).

20. L. E. Ibañez, Phys. Lett. 126B, 196 (1983); B. A. Ovrut and S. Raby, Phys. Lett. 154B, 130 (1985), 138B, 72 (1985).

Footnote

F1. This analysis demonstrates that, given a GUT scale at $\sim 10^{15}$ Gev, dimension 5 operators must be suppressed. If we would have considered the d = 5 operators first, we would have concluded that $M \geqq 10^{32}$ Gev which is much larger than M_{pl} and thus unnatural.

DISCUSSION

CHAIRMAN: S. Raby

Scientific Secretaries: Z. Bern, J.E. Bjorkman and G. D'Ambrosio

– *Y. Shamir:*

Are there GUT models or cosmological models in which CP violation is dynamical and not put in by hand? Is this connected to matter–antimatter asymmetry?

– *S. Raby:*

Yes, there are. As an example, recently a model was discussed by Lawrence Hall and his collaborators in the contest of supersymmetric GUT. In the model the Lagrangian is CP invariant, but the vacuum is not. As a result there are CP violating phases in scalar field vacuum expectations values which feed into the low energy theory via scalar quark mass matrices. These in turn, feed into $K^0 - \overline{K}^0$ mixing in a standard way, thus contributing to CP violation. It is possible for this CP violation to be connected to the observed matter–antimatter asymmetry. In the contest of standard baryogenesis scenario, matter–antimatter asymmetry is generated through the out of equilibrium decays of heavy particles. In order to generate an asymetry, the decay process must also violate the symmetries C, CP and baryon number.

– *Y. Shamir:*

In the grand unified $SU(5)$ Higgs multiplet, you discussed, how can one give the Higgs triplet a large mass and give the Higgs doublet a small mass without resorting to fine tuning the mass matrix. Could you explain the method using multiply connected manifolds.

– *S. Raby:*

This was partly discussed by Prof. Gross in his lecture in connection with compactification on Calabi–Yau manifolds. In a multiply connected manifold you can have a non–vanishing gauge potential with vanishing field strength. It will give a contribution to a Wilson loop measuring the flux through the hole in the compactified space. Such gauge field configurations can be used to break the gauge group, say E_6, down to some smaller group. This scenario was discussed by Witten, Breit, Ovrut, Segrè and others.

– *K. Meissner:*

Can the experimental lower bound on the proton lifetime distinguish between supersymmetric and nonsupersymmetric grand unification models or can, modifying the parameters in both models, give us any proton lifetime we wish?

– *S. Raby:*

In the minimal $SU(5)$ GUT one can obtain a prediction for the proton lifetime of order $10^{28\pm2}$ years and for the dominant decay mode $p \rightarrow \pi^0 e^+$. A search for proton decay in this channel now finds a lower bound on the lifetime of order 10^{32} years. Thus the minimal model is ruled out experimentally. The minimal supersymmetric $SU(5)$ GUT, on the other hand, cannot make a definite prediction for the lifetime since (a) it depends on the masses and couplings of particles in the low energy theory which have yet to be observed and whose values are not restricted by any suymmetry and (b) it depends on the mass and couplings of color triplet Higgs fermions which, again, are not restricted by symmetries. The model does predict, however, that the dominant decay mode is $p \rightarrow K^+ \overline{\nu}_\mu$.

In either supersymmetric or nonsupersymmetric theories these predictions can be changed by altering the models. For example, in $SU(5)$ one assumes that irreducible representations of $SU(5)$ are diagonal in generation number, i.e. a $10 = (Q, \overline{U}, \overline{E})$ contains only a single generation. If the 10 contained two different generations (say $Q = (u, d), \overline{U} = \overline{t}$) then there would be no proton decay at tree level. Reiss and Rudaz have shown, in a recent paper, that non–trivial fermion mass matrices can lead to sufficient generation mixing in the gauge multiplets so as to suppress the proton decay rate and/or lead to new dominant decay modes.

Predictions of the proton lifetime are also affected by the light states in the model. As shown by Frampton and Glashow, $SU(5)$ can be made consistent with experiment with a non–minimal light spectrum of fermions.

Finally, larger grand unified gauge groups with more breaking patterns and scales will affect the prediction for the proton lifetime.

– *P. Shotton:*

What is the present theoretical opinion concerning magnetic monopole catalyzed proton decay?

– *S. Raby:*

The present status is that people still believe that monopoles catalyze proton decay. There is no question of this. The problem is in experimentally observing it. The idea of monopole catalysis of proton decay is being used for monopole searches. In one such experiment a huge swimming pool of water is used to look for Cherenkov radiation from proton decay. If a monopole passes through the water one expects to see a trial of proton decays along the monopole trajectory. The absence of such events places a strong upper bound on the monopole flux. Unfortunately monopoles have yet to be discovered.

– *G.D. Coughlan:*

You mentioned that strange decay products seem to be favoured by dimension five proton decay operators. Is it not possible for non– strange modes to dominate for a reasonable range of scalar masses and mixing angles? In particular, is it not possible for a heavy top squark in the loop to more than compensate for Cabibbo suppression?

– *S. Raby:*

Yes, there are many possibilities depending on parameter values. For example, in some range of parameters graphs involving gluino exchange dominate, while in another range those involving wino exchange dominate.

The gluino graphs were shown by Belayev and Vysotski to cancel if the up, down and strange squarks had the same mass. However, Multinovic, Pal and Senjanovic pointed out that since the squark masses are actually split, gluino exchange graphs can be significant. They may in fact dominate as shown by Chadha, Daniel, Coughlan and Ross, for a top squark mass as low as 40 GeV and large top mixing angles, leading to an enhanced $p \to K^0 \mu^+$ decay rate.

Finally, Arnowitt, Chamseddine and Nath have shown that for a certain range of parameters cancellation can occur in the strange quark decay mode in wino exchange graphs. In these graphs, contributions from charm and top squarks in the loop cancel and consequently, the down quark decay mode is dominant. This leads to $p \to \pi^+ \bar{\nu}$ and $n \to \pi^0 \bar{\nu}$ or $\eta \bar{\nu}$ as the dominant decay modes.

– *M. McGuigan:*

Can experimental limits on lepton number violating processes provide any limits on the coefficients of your effective Hamiltonian?

– *S. Raby:*

I do not know any definite limit on that. The most stringent limit may come from the process $\mu \to e\gamma$, but I do not know of anyone that has checked that. For those operators where $\Delta B = \Delta L$, I assume the most stringent limit comes instead from baryon number violating processes.

– *S. Giddings:*

You used the dimension 6 operators to set the symmetry breaking scale, and then you had to argue that the coefficients of the dimension 4 and 5 operators had to be vanishingly small. Why did you not start with,say, the dimension 5 operators rather than the dimension 6 operators?

– *S. Raby:*

The only reason that I did not do that is that you would probably get a mass scale of order 10^{25} GeV. It would just be unreasonable since (a) the renormalization group equations (which really do determine this scale) give 10^{16} GeV or so and (b) since this scale is much larger than the Planck scale its significance would be highly suspect. Therefore, we really do need vanishingly small coefficients.

– *M. Quiros:*

The amplitude for $p \to K^+ \bar{\nu}_\mu$ is proportional to $1/sin2\theta_h$ where $tan\theta_h \equiv v/\bar{v}$, so it seems that $A \to \infty$ in the limit $\bar{v} \to 0$ ($m_d, m_e \to 0$). Do you have an explanation for it?

– *S. Raby:*

Really what we have is $\lambda_c = \frac{m_c}{v}$ and $\lambda_s = \frac{m_s}{\bar{v}}$. The amplitude is proportional to

$$A \approx \lambda_c \lambda_s = \frac{m_c m_s}{sin2\theta_h} \frac{g_2^2}{2m_W^2}$$

where

$$sin2\theta_h = \frac{v\bar{v}}{(v^2 + \bar{v}^2)}$$

and

$$m_W^2 = \frac{g_2^2(v^2 + \bar{v}^2)}{2}$$

which is well behaved for $\bar{v} \to 0$ (λ_c and λ_s held constant).

– *M. Quiros:*

What are the predictions for the proton decay in superstring inspired model?

– *S. Raby:*

Superstring inspired models generally contain light 27 dimensional representations of E_6. In the breaking pattern $E_6 \to SO(10) \to SU(5)$, the irreducible representation 27 breaks down as follows: $27 \to 16+10+1 \to (10+\bar{5}+1)+(5+\bar{5})+1$. The 5 and $\bar{5}$ coming from the 10 of $SO(10)$ contain dangerous light color triplet fields which can mediate rapid proton decay. We must thus avoid proton decay from dimension 4 operators which contain these fields. This can be accomplished either by discrete symmetries which forbid the dangerous dimension 4 operators or by finding a way to give the color triplet states masses greater than about 10^{17} GeV. This problem has yet to be solved.

– *G. Miele:*

How do you solve the problem of the vacuum degeneracy of scalar VEV's using soft SUSY breaking?

– *S. Raby:*

Vacuum degeneracy is a common feature of SUSY theories, which is typically lifted once SUSY is spontaneously broken. For example, if SUSY is broken at a scale Λ^2 of order $m_W M_{pl}$ in some hidden sector, as is commonly believed to occur in supergravity models, then soft SUSY breaking operators will appear in the effective low energy theory at the tree level. The SUSY breaking scale in the effective low theory is fixed by the gravitino mass $m_g \cong \Lambda^2/M_{pl}$. These soft SUSY breaking operators lift any vacuum degeneracy. They are renormalizable and they do not contribute to any radiative quadratic mass divergences for scalars. The types of terms which are allowed are as follows:

$$\mathcal{L}_{soft\ SUSY\ breaking} \simeq \frac{M_i}{2}(\lambda\lambda)_i + AW_3\Big|_{\theta=0} + BW_2\Big|_{\theta=0} - \widetilde{m}_a^2 \sum_{all\ a} \Big|\phi_a\Big|^2 \qquad i = 1, 2, 3$$

where λ_i are gaugino fields, $W_n, n = 2, 3$ are the quadratic and cubic terms in the superspace potential, and ϕ_a represent all the scalars in the theory. The quadratic mass term for the ϕ_a breaks the degeneracy. Note, if $\widetilde{m}_a^2$ is negative, then one has to be careful not to have a potential which is unbounded from below.

G. Marx:

How do you solve the problem of the [illegible]

[illegible]:

Vacuum degeneracy is a common feature of SUSY theories, which is typically lifted once SUSY is spontaneously broken. [illegible] are as follows.

[illegible]

[illegible]

SUPERSTRING PHENOMENOLOGY: AN OVERVIEW*

Michael Dine

Physics Department
City College of the City University of New York
New York, NY 10031
and
The Institute for Advanced Study
Princeton, New Jersey 08540

ABSTRACT

Elements of superstring phenomenology, as it presently exists, are reviewed. The coupling constants of string theory are identified. Classical solutions should be relevant at weak coupling; the connection to conformally invariant non-linear sigma models is explained. Compactifications on Calabi-Yau spaces and orbifolds which preserve N=1 supersymmetry are argued to be the most promising. Rather simple phenomenological considerations are shown to severely constrain the properties of the compact six dimensional space. The cosmological constant and the dilaton potential are discussed and we explain why, at weak coupling, there is probably no ground state which resembles our world.

INTRODUCTION AND OUTLINE

Superstrings appear to provide consistent quantum theories of gravity and gauge interactions[1,2]. This by itself would make them worthy objects of study. What is still more remarkable is that, even with our primitive understanding of these theories, we can go a long way towards developing a superstring phenomenology. The present article seeks to provide an overview of the enormous amount of work which has been done on the subject. I hope to make clear the set of principles

* Lectures at the 1986 International School of Subnuclear Physics, Erice, Sicily

and assumptions which guide present efforts. Perhaps the most important of these is the assumption of weak coupling. In order to do phenomenology in string theory, one must assume not only that string theory describes nature, but that we are also in a perturbative regime, in the sense that corrections involving string loops are small. After all, everything we currently know about string theory comes from analyses of perturbation theory. Whether or not it is stated explicitly, this hypothesis underlies all current efforts to connect string theory with observations. Within this framework there have been some striking successes, and candidate ground states with many features which resemble our world have been found. From this work, plausible scenarios for how string theory might make contact with low energy physics have emerged. But some serious concerns, if not obstacles, have arisen as well. Clearly, then, if we are to comprehend these recent developments and make further progress, it is important, first, to understand what the coupling constant of string theory is. In the second part of this article, we will see that the coupling constant is determined by the expectation value of a dynamical field[3]. This expectation value is classically undetermined. The effective coupling of string theory can only be determined by a quantum, dynamical, calculation.

If we assume that the coupling is weak, it is appropriate (as in conventional field theory) to examine classical solutions of string theory. In the third section, we will explain what it means to find a classical solution of string theory[4], and we will see that vast numbers of such solutions are known. We will describe a subset of these, Calabi-Yau spaces[4] and compactifications on orbifolds[5], which exhibit promising features: space-time is four dimensional, the cosmological constant vanishes classically and (it is believed) in perturbation theory, the low energy gauge group is something like SU(3)×SU(2)×U(1), there is a chiral structure of generations, and the low energy theory is supersymmetric. We will discuss the various ways in which the features of these solutions may be explored and enumerate some of these features. In the fourth section, we will consider the low energy phenomenology of these compactifications in more detail. We will see that there are a number of critical tests which rule out most (but fortunately not all!) candidate vacua. These include proton stability, neutrino masses, and renormalization group considerations. Finally, we will return to question the underlying assumptions of all this work. We will review the cosmological constant problem and see that, unfortunately, superstring theory has to date shed no new light on this fundamental question. Related to this problem we will find another: if superstring theory describes nature, it is almost certainly strongly coupled[6]. For these questions, we will have no easy answers. For definiteness we will focus throughout on the $E_8 \times E_8$ heterotic string.[2]

COUPLINGS AND SCALES

String theories are often described as having no (dimensionless) parameters. Before launching into any explanation of possible models, it is important to understand what this statement means and what the parameters of string theory are. We can view the string tension, α', as providing a fundamental scale for string theory. We will define $M_s = (\alpha')^{-1/2}$, to be the string scale. M_s is *not* a priori the ten-dimensional Planck mass, or the four-dimensional one. In addition, vertex operators for emission of particles such as gravitons, gauge bosons, etc., involve a dimensionful coupling constant called g; for the heterotic string there is only one such quantity.

For sufficiently low energies, and for sufficiently weak coupling (a notion which we will define a posteriori) it should be possible to "integrate out" all the massive modes of the string, and describe the light modes - the graviton, $g_{\mu\nu}$, antisymmetric tensor, $B_{\mu\nu}$, gauge bosons A_μ, the dilaton, D and their supersymmetry partners - by an effective Lagrangian. This Lagrangian will contain operators of arbitrarily high dimensions and will have associated with it a cutoff of order M_s. It should possess all of the symmetries of the full theory: general coordinate invariance, gauge invariance, supersymmetry, and the gauge symmetry associated with $B_{\mu\nu}$ ($B_{\mu\nu} \rightarrow B_{\mu\nu} + \partial_\mu \epsilon_\nu - \partial_\nu \epsilon_\mu$). The terms involving at most two derivatives of the field follow from supersymmetry alone[7]. In one convenient parametrization of the fields[8],

$$L = \phi^{-2}\left(-\frac{1}{2}R - \frac{1}{4}F^2 - \frac{3}{4}H^2 + 2\left(\frac{\partial\phi}{\phi}\right)^2 + \text{fermionic terms}\right) \tag{1}$$

where $\phi = e^{-D}$, and we have set $M_s = 1$. Here R is the usual curvature tensor, $F_{\mu\nu}$ the Yang-Mills field strength, and

$$H_{\mu\nu} = \partial_\mu B_{\mu\nu} + Chern - Simons \quad terms. \tag{2}$$

The main point to note is the factor ϕ^{-2} out front. At tree level, this is a feature not only of the lowest derivative terms but also of all the higher dimension terms as well[9]. This fact is easily derived in string theory[10].

Now in field theory, a factor which multiplies the whole Lagrangian is a coupling constant (for example, the QCD Lagrangian is frequently written as $-\frac{1}{4g^2}F_{\mu\nu}^2$). So, if we expand the dilaton about its vacuum expectation value, $D = D_o + \delta D, \phi_o = e^{-D_o}$ is the coupling constant of string theory. In particular string loops come with factors of ϕ_o^2(ϕ_o here has been defined to be dimensionless). It is important in making this argument that M_s is the only scale appearing here, as well as the cutoff. Otherwise, dimensionless ratios of such scales could appear in loops as well.

If we examine the Lagrangian of Eq. 1, it is clear that if we expand all of the fields about ten-dimensional flat space; there is no potential for D. So D_o is undetermined. The higher derivative terms do not change this. In field theory, in such a situation, any choice of D_o is as good as any other as a vacuum for the theory (there are no transitions from one such vacuum to another; see, e.g., Ref. 11). The same is true in string theory. So, at the classical level, we have encountered an infinite set of vacua. In fact, this is almost certainly an *exact* property of string theory in ten dimensions, at least at weak coupling; supersymmetry would appear to forbid any operator in the low energy Lagrangian which would lift the degeneracy. The strength of the coupling depends on the vacuum we choose. These flat ten-dimensional states are almost certainly *exact* (perturbative and non-perturbative) ground states of string theory[12]!

CLASSICAL SOLUTIONS OF STRING THEORY

Fortunately, these are not the only classical solutions of string theory. As we will now discuss, there exist many other classical solutions, some of which rather closely resemble our world. Clearly we would like six dimensions to be compact, while four are flat. If for the moment, we suppose the internal space is large compared to M_s, we don't expect the massive modes of the string to be highly excited, so that it should be possible to focus on the light fields. A bosonic string propagating in a background metric, $G_{\mu\nu}$, is described by the action (in conformal gauge).

$$S = \frac{1}{4\pi\alpha'} \int d^2\sigma G_{\mu\nu}(X)\partial_\alpha X^\mu \partial_\alpha X^\nu \; . \tag{3}$$

If $G_{\mu\nu}$ is some complicated function of position, this is the action for a highly non-trivial two dimensional field theory – a non-linear sigma model.

In the case of the heterotic string, in addition to a background metric, we may have non-trivial background gauge, antisymmetric tensor, and dilaton fields. Perturbation theory is developed in this sigma model by considering a reference metric, $G^\circ_{\mu\nu}$, such that the volume of the internal space, measured in string units, is of order one. If we write $G_{\mu\nu} = R^2 G^\circ_{\mu\nu}$, R^{-2} is the sigma model coupling. Working in the fermionic formulation of the heterotic string, we have, in addition to the ten bosonic fields, X^μ, ten right moving two dimensional spinors, $\psi_\mu(\tau - \sigma)$, and 32 left moving spinors, λ^A and λ^a, $A = 1, \ldots, 16$ and $a = 1, \ldots, 16$. These latter transform as a vector representation of an $O(16) \times O(16)$ subgroup of $E_8 \times E_8$.

In the presence of background fields, the action looks like[13,14] (for the moment we take $B_{\mu\nu}$=0, and ignore the dilaton field):

$$S = \frac{1}{4\pi\alpha'} \cdot \int d^2\sigma \{ G_{\mu\nu} (\partial_\alpha X^\mu \partial_\alpha X^\nu + \psi^\mu D_- \psi^\nu) + \lambda^A D_+ \lambda^A + \lambda^a D_+ \lambda^a + F^{AB}_{\mu\nu} \lambda^A \lambda^B \psi^\mu \psi^\nu + F^{ab}_{\mu\nu} \lambda^a \lambda^b \psi^\mu \psi^\nu \} \quad . \tag{4}$$

Here D_+ is the covariant derivative with respect to the background gauge field, D_- the covariant derivative with respect to the background metric, and $F^{AB}_{\mu\nu}$ and $F^{ab}_{\mu\nu}$ the background Yang-Mills field strengths. At the classical level, we are not interested in just any backgrounds, but in those which represent solutions of the string equations of motion. Since no one, at present, actually writes down and solves string equations of motion (except for trivial backgrounds) we have to infer that a given configuration is a solution of the equations of motion more indirectly. The key here is conformal invariance. This is not the place to review conformal field theory[15], but a simple argument[4] shows that if the two-dimensional field theory of Eq. 4 is conformally invariant, then the string configuration with these expectation values for the massless fields is a solution of the string equations of motion. To see this, it is perhaps easiest to employ the Polyakov program[16,17]. There, in flat space, scattering amplitudes are computed by calculating in a path integral expectation values of strings of "vertex operators",

$$\int d^2\sigma \sqrt{g} V_i(k, X^\mu), \tag{5}$$

where g is the world-sheet metric and k is the momentum. In order that this object be invariant under Weyl rescalings ($g_{\mu\nu} \rightarrow e^\phi g_{\mu\nu}$) we require that V_i have dimension two (in a complex basis, (1,1)). This requirement carries over to the case of non-trivial backgrounds: we must find the operators of dimension two in the appropriate non-linear sigma model. The corresponding particles will have various masses, $m_i^2 = k_i^2$, where k_i is the four-dimensional momentum; of particular interest are those which are massless.

If some configuration is not a solution of the string equations, then the massless (from the point of view of four dimensions) particles will have tadpoles, i.e. non-vanishing one-point functions. Since for massless particles, zero momentum coincides with the mass shell, we can try to compute such tadpoles by evaluating the expectation value of a single vertex operator.[4] But such a vev necessarily vanishes. To see this, we use the non-anomalous SL(2,C) invariance of the theory. Under this,

$$z \rightarrow z' = \frac{az+b}{cz+d} \qquad ad - bc = 1 \tag{6}$$

$$V \rightarrow \left| \frac{\partial z'}{\partial z} \right|^2 V = | cz + d |^4 V \quad . \tag{7}$$

So, using the SL(2,C) invariance of the sigma model vacuum, taking $b = 0$,

$$< V(0) >=| d |^4 < V(0) >= 0 \tag{8}$$

and there are no tadpoles. To each conformally invariant σ-model corresponds a classical solution of string theory.

For the heterotic string, a particularly interesting set of compactifications were found by Candelas et al.[9] These authors noted that, from explicit computations in supersymmetric non-linear sigma models, the requirements for conformal invariance could be satisfied in the following way. Choose for the internal space a complex space with a Kahler metric. This means that the six real coordinates of the internal space can be grouped into three complex coordinates, x^1, x^2, x^3 and $\bar{x}^1, \bar{x}^2, \bar{x}^3 (\bar{x}^i = x^{i*}$, etc), and that the metric takes the form

$$g_{i\bar{i}} = \partial_i \bar{\partial}_i K \tag{9}$$

where K is a function of x_i and $x_{\bar{i}}$ called the Kahler potential (for a simple introduction to sigma models with Kahler geometry, see for example, Ref. 18). Choose a metric such that the Ricci tensor vanishes. For a Kahler manifold,

$$R_{i\bar{i}} = -\partial_i \partial_{\bar{i}}\ ln\ det\ g_{i\bar{i}}\ . \tag{10}$$

A theorem of Calabi and Yau guarantees that such a metric can always be found.

Our six dimensional space locally has tangent space group $O(6)$. $O(6)$ is isomorphic to SU(4). When we describe the coupling of spinor fields to gravity, it is necessary to introduce an object called the spin connection, $\omega^a_{\mu b}$.[19] Here μ is a conventional vector index; a and b refer to the six dimensional tangent space. We go from one index to the other by means of the vierbein (a terminology which refers to four dimensions but which I will use anyway) e^a_μ, e^μ_a,

$$e^a_\mu e^a_\nu = g_{\mu\nu} \quad ; \quad e^a_\mu e^b_\nu g^{\mu\nu} = \eta^{ab}\ . \tag{11}$$

In terms of ω_μ,

$$R^{ab}_{\mu\nu} = e^{\rho a} e^{\sigma b} R_{\mu\nu\rho\sigma} \tag{12}$$

looks formally like a Yang-Mills field strength. As an O(6)=SU(4) matrix,

$$R_{\mu\nu} = \partial_\mu \omega_\nu - \partial_\nu \omega_\mu + [\omega_\mu, \omega_\nu] \quad . \tag{13}$$

The statement that the Ricci tensor vanishes, one can easily show in this language, means that $R_{\mu\nu}$ in Eq. 13, acting on three complex coordinates, is traceless. In particular, it is an SU(3) matrix. ω_μ can be taken to be an SU(3) matrix as well (thus the phrase "SU(3) holonomy").

In the case of the heterotic string, we must also specify a background gauge field. The authors of Ref. 4 made a simple choice. They noted that E_8 has an SU(3) subgroup. They suggested one take the gauge fields in this SU(3) to be non-vanishing, and to be precisely equal to $\omega^i_{\mu j}$ (using complex coordinates). Then $F_{\mu\nu} = R_{\mu\nu}$ as SU(3) matrices. The connection which appears in Eq. 4 is now the same for both left-moving and right-moving fermions, so the model has a left-right symmetry. In fact, the model possesses two left-moving and two right moving supersymmetries, and is said to have (2,2) supersymmetry.[20] For such models, the β-function has been computed through four loop order.[21,22,23] To three loop order, the model is conformally invariant as it stands. At four loop order, a conformally invariant model may be found from this one perturbatively in α'/R^2.[24] In fact, such a solution may be constructed to all orders in perturbation theory.[24] In Ref. 25, it was argued that this is true even non-perturbatively in the sigma model. We will come back to these issues later.

What does this vacuum look like, if viewed not from two or ten, but from four dimensions? First, the low energy theory has N=1 supersymmetry in four dimensions. Perhaps the easiest way to see this is to exhibit the gravitino. To do this, note that if we choose a gauge where (I denotes a generic internal index; a,b are SU(4) indices)

$$\omega_{I\ b}^{a} = \begin{pmatrix} & & \vdots & 0 \\ & \omega_I & \vdots & 0 \\ \cdots & \cdots & \vdots & 0 \\ 0 & 0 & 0 & 0 \end{pmatrix}, \tag{14}$$

then the constant spinor on the internal space,

$$\epsilon_a = \begin{pmatrix} 0 \\ 0 \\ 0 \\ \epsilon \end{pmatrix}, \tag{15}$$

is covariantly constant. The spin-3/2 field has indices $\psi^\mu_{a\alpha}$, where μ is now a 4-dimensional Minkowski index, α a four dimensional (two-component) spinor index, and a an SU(4) index. Take

$$\psi^\mu_{a\alpha} = \psi^\mu_\alpha(x^\nu)\epsilon_a(y^I) \tag{16}$$

where x^μ refers to Minkowski coordinates and y^I to the internal coordinates. I leave it as an exercise to check that if ψ^μ satisfies the ten-dimensional massless Rarita-Schwinger equation, $\tilde{\psi}^\mu$ satisfies the massless four dimensional Rarita-Schwinger equation.

We must also determine the low energy gauge symmetry. Apart from the other E_8, which is so far untouched, this is just E_6, which is the subgroup of E_8 which commutes with SU(3). In fact, this E_6 (as well as the other E_8) may be broken further by Wilson lines to groups like SU(3)×SU(2)×U(1)^3.[4,26] Before explaining this, let me discuss the low energy spectrum at this stage. In addition to the massless particles so far described, there are many others. Clearly there are gravitons and gauge bosons, whose internal wave functions are just constants. In other words, we expand the metric in harmonics on the internal space

$$g_{\mu\nu}(x,y) = \sum g_{\mu\nu}^{(n)}(x)\phi^{(n)}(y) . \tag{17}$$

The coefficient of the constant function is viewed, in four dimensions as a massless particle.

The equation $R_{IJ} = 0$ is clearly unaffected by multiplying the metric everywhere in space-time by a constant. Correspondingly, the size of the internal space is classically undetermined, and there is a massless field in four dimensions, which we can call $R(x)$, whose internal wave function is just proportional to the internal metric. In fact, the "shape" of the internal space is not fully determined either. The fields which describe the shape are referred to as the "moduli". In addition to these, there are other massless particles associated with the structure of the internal space.[4]

Perhaps more interesting are the fields which are charged under the gauge group. These come from the gauge fields and their superpartners, which lie in the 248 of E_8. Under SU(3)×E_6 the 248 decomposes as

$$248 = (78,1) + (1,8) + (\bar{3},27) + (3,\overline{27}) . \tag{18}$$

The matter field will lie in 27's of E_6; they will come in supersymmetry multiplets (complex scalars and chiral fermions). The 27 is a "nice" representation: it contains all the particles of a standard generation, plus some extras. From the point of view of SU(5), for example,

$$27 = \bar{5} + 10 + 5 + \bar{5} + 1 + 1 . \tag{19}$$

The wave functions of these zero modes are more complicated, but the number of massless 27's and $\overline{27}$'s are determined by topological considerations. In particular, the number of 27's minus the number of $\overline{27}$'s, which is the number of generations, turns out to be the Euler number divided by two. Examples of Calabi-Yau spaces with three generations are known.

This is not the place for a complete review of the Wilson line mechanism for gauge symmetry breaking. I strongly recommend reading Witten's article.[26] Let me just note that, in general, the manifolds we are considering are not simply connected. So there exist closed curves which cannot be shrunk to points. As we go around such a curve, call it C, the vector potential must be single valued. However, it is possible that

$$A_I = g^{-1}\partial_I g \tag{20}$$

where g is not single valued. Then the Wilson line

$$\begin{aligned} U &= Pe^{i\oint dx_\mu A^\mu} \\ &= g^{-1}(1)g(0) \ . \end{aligned} \tag{21}$$

Here P denotes path ordering, g(0) is the initial value of g, and g(1) the value after one traversal of the loop. Since the field strength vanishes for such a configuration, there is no energy cost to such a vacuum expectation value for the gauge field. Thus we have an additional set of labels for candidate vacua, the expectation values for the matrices U. As mentioned above, these break the gauge group further. The number of generations remains the same, but the numbers of massless 27's and $\overline{27}$'s, may be different (we may think loosely of fields from 27's and $\overline{27}$'s pairing to gain mass). Of course, since E_6 is broken, complete 27's need not gain mass. In fact, it is possible to obtain massless weak doublets without color triplet partners.[26,27] This is extremely important for it means we may be able to obtain Higgs fields to break SU(2)xU(1) and give mass to quarks and leptons, without obtaining light color triplets which would mediate rapid proton decay. We should also stress, at this stage, that at tree level the SU(3)×SU(2)×U(1)× ... gauge couplings are equal at the unification scale, while the Yukawa couplings need not obey the relations implied by E_6.[26]

In addition to knowing the massless spectrum, one would like to compute the couplings among the various fields. In particular, one would like to derive a low energy effective Lagrangian from which massive modes (both higher string modes and higher Kaluza-Klein modes) have been integrated out. This can be done at tree level by computing S matrix elements at low energy using appropriate vertex operators, and then writing down a Lagrangian which at tree level reproduces these. Of particular interest are Yukawa couplings of the various E_6 non-singlet fields. In practice, these have been calculated by a somewhat different method.[28] Starting from ten dimensions, with the coupling

$$\int d^{10}x\bar{\psi}A\psi$$

replacing A_I and ψ by appropriate zero-mode wave functions and integrating over the internal space, we obtain effective four dimensional couplings, which can be related to topological quantities. Interestingly, one typically finds that many couplings vanish.

Some care must be taken in applying these results, however. In particular, one must compute the normalization of the kinetic terms for the various fields so the Yukawa couplings computed in this way can be rescaled to obtain the usual ones[43]. Also, one might expect corrections to these computations coming from higher dimension operators in the effective ten-dimensional theory, or equivalently from higher order terms in the σ-model.

In fact, there is an interesting non-renormalization theorem for σ-model perturbation theory due to Witten.[25] A four-dimensional, N=1 supergravity Lagrangian (such as would describe the interactions of the fields here) is specified (for terms with two derivatives or less) by giving three functions of the various chiral fields.[30] In particular, potentials, masses and Yukawa couplings for matter fields are obtained from an analytic function of the chiral fields called a superpotential. Witten shows, by a remarkably simple argument (which I don't have time to repeat here) that this function is not renormalized in σ-model perturbation theory (even taking into account the shifts in the background fields necessary at four loops and beyond). This is quite a powerful result. It guarantees, for example, that fields which are "accidentally" massless at lowest order (such as the would-be Higgs doublets described above) are massless to all orders in α'. It also means that the computation of Yukawa couplings of Ref. 28 is in some sense exact, though the kinetic terms will be corrected in each order.

The non-renormalization theorem is also the basis of a proof, due to Witten[29], that these configurations are solutions to the classical string equations of motion to all orders in perturbation theory. The idea is very simple. If, at some order, the configuration of interest ceased to be a solution, this would show up as a tadpole for some light field. However, for such a tadpole to appear the superpotential would have to be renormalized. But this is forbidden. This theorem also shows that the ground state is supersymmetric. Note that this proof does not preclude the possibility of shifts of heavy fields, corresponding to changes in the background metric, gauge fields, etc. Thus it is not inconsistent with the result of Ref. 22-24.

In fact, this theorem was the starting point for the construction of new solutions of the string equations.[29,31] These solutions do not have $\omega_\mu = A_\mu$. The corresponding sigma models, as a result, are not left-right symmetric and have 2 right-moving and zero left moving world-sheet supersymmetries, (2,0) supersymmetry.[20] The resulting models look like unified theories with gauge group O(10) or SU(5). This is desirable, because the number of light charged particles is smaller (the extra 5 and

5 of Eq. 19 are gone). These models were, again, solution of the string equations to all orders in $\alpha\prime$.

However, we know that there are effects in these models which are smaller than any power of α'/R^2. Instantons, in particular give contributions to amplitudes of order $e^{-aR^2/\alpha\prime}$, where a is some constant. It has recently been shown that instantons, in fact, do renormalize the superpotential.[25] This raises several questions and concerns: (1) Are these models still solutions of the string equations? (2) Are the various "accidentally" massless particles still massless? (3) What about corrections to Yukawa couplings?

The answers to these questions have also been provided in Ref. 25. It turns out that, in general, the (2,0) theories are not solutions of the string equations (though there may be exceptional cases). On the other hand, the (2,2) theories appear to be exact solutions even non-perturbatively. The extra massless particles in the 27 and $\overline{27}$ representations - our candidate Higgs particles - appear to be exactly massless. Certain other "accidentals", however (such as those from the SU(3) octet discussed in Ref. 29) do gain mass. Yukawa couplings are corrected. Couplings which vanish at lowest order for reasons not connected with symmetries are expected to receive non-zero contribution from instantons.

So far, our analysis has been based on the assumption that the internal space is large compared to M_s^{-1}. In fact, even if string theory is weakly coupled, it is easy to show that the compactification scale must be larger than the string scale, i.e. $R < M_s^{-1}$.[8,32] Thus the effects which are formally small here are not really small. Any coupling which is non-vanishing is likely to be of order one.

You may be asking: what about corrections from the string perturbation expansion (higher genus surfaces) to these results. If this were ordinary field theory, we would not have to worry, at least in perturbation theory. There, non-renormalization theorems would prevent further corrections to the superpotential from loop graphs. For string theory, this is also believed to be the case, and there has been some progress in proving this.[33]

There exists another very interesting class of compactifications of string theory: compactifications on orbifolds[5]. These can frequently be thought of as limiting cases of Calabi-Yau spaces. They are also solutions of the classical string equations. Unlike Calabi-Yau spaces, the metrics are known and simple (flat), and the spectrum and interactions may be calculated explicitly. The idea of the orbifold construction is best illustrated by an example, the so-called "Z orbifold." One starts by taking the compact space to be three orthogonal copies of the torus of fig. 1. The angle θ is 60°, and points on opposite sides of the torus are identified,

$$\begin{aligned} x_1 &= x_1 + R_1 \\ x_2 &= x_2 + R_2 cos\theta \end{aligned} \tag{22}$$

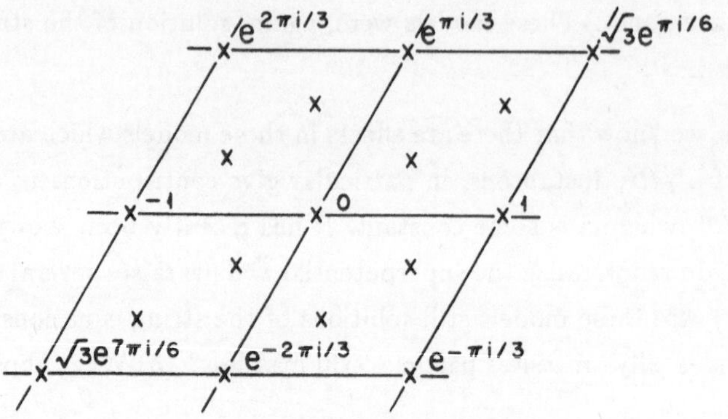

Figure 1. Torus for Z orbifold

In addition to this identification, we also identify points under rotations by $\frac{\pi}{3}$. It is easy to see that there are 27 points which are invariant under this transformation. Using complex coordinates,

$$\begin{aligned} z^1 &= x^1 + ix^2 \\ z^2 &= x^3 + ix^4 \\ z^3 &= x^5 + ix^6 \end{aligned} \tag{23}$$

these have

$$z^i = 0, \frac{1}{\sqrt{3}}e^{i\pi/6}, \frac{2}{\sqrt{3}}e^{i\pi/6}\,. \tag{24}$$

Here, by invariant, we mean up to shifts by the vectors which define the torus ("lattice vectors"). This space is flat everywhere, but is rather singular. For, if we parallel transport a vector through an angle $\frac{\pi}{3}$ about a fixed point, we come back to our starting point, but with the vector only rotated by $\frac{\pi}{3}$. Such a point is called a conical singularity, and the angle by which the vector is rotated (or rather 2π minus this angle) is called the deficit angle. We can think of there being an infinite amount of curvature concentrated at the fixed point. This manifold has SU(3) holonomy, since the change in any vector upon parallel transport about a closed curve is an SU(3) transformation. (Gauge field theorists may want to think of this in terms of spin connections and SU(3)-valued Wilson lines). Just as for Calabi-Yau spaces, the low energy theory is supersymmetric, since there is a covariantly constant spinor (here it is quite easy to construct the massless spin-$\frac{3}{2}$ state explicitly).

In string theory, we must do two-dimensional field theory on this space. Essentially, we need to enumerate the allowed boundary conditions for our two dimensional fields, perform normal mode expansions for each possible boundary condition, and compute the spectrum and interactions as for strings in flat ten dimensions. In the heterotic string theory, in addition to the boundary conditions for the "space-time" fields (the bosonic string coordinates and the right-moving Green-Schwarz fermions) we must also discuss the boundary conditions for the "gauge" excitations, for example for the 32 left-moving spinors of the fermionic formulation. In the fermionic formulation, only an O(16)×O(16) symmetry is manifest. The analog of embedding the spin connection in the gauge group is the following. Take six fermions from one O(16). There is an obvious O(6) group which acts on these, and this O(6) has an obvious SU(3) subgroup. This SU(3) will be identified with the SU(3) holonomy group which acts on space-time degrees of freedom, and in particular on the Green-Schwarz fermions. These eight fermions decompose under SU(3) as a $3 + \bar{3} + 1 + 1$. (The eight RNS fermions of light-cone gauge have an identical decomposition). The six gauge fermions transform as a $3 + \bar{3}$ under SU(3). Identifying the gauge field with the spin connection here just means that we treat the left-moving and right moving 3's and $\bar{3}$'s identically. This leads to theories with left-right symmetry from a two-dimensional viewpoint, i.e. (2,2) models. Other choices are possible, leading to (2,0) models. All of these models are exactly conformally invariant and it is possible to compute scattering amplitudes explicitly.[34] In particular, the results of Ref. 25 have now been verified by orbifold computations. The phenomenology of orbifolds has only been partially explored. This is clearly an area where further work is necessary.

LOW ENERGY PHENOMENOLOGY

We have exhibited an enormous vacuum degeneracy at the classical level. Hopefully, some of this degeneracy is lifted quantum mechanically and a ground state which resembles our world is in some way selected. We will explore this question further in the next section, where we will discuss supersymmetry breaking. For the moment, we can attempt, instead, to see if there is among these vacua one which *does* in fact resemble our world. We will focus, exclusively, on vacua which have N=1 supersymmetry at the classical level. Vacua with more supersymmetry ($N \geq 1$) will not have chiral fermions. Vacua with no supersymmetry, as we will discuss in the next section, are already unacceptable at the one loop level. In any case, low energy supersymmetry is a property we might like to have in order to understand the large hierarchy between the Planck scale, M_p, and the weak scale, M_W.[35]

To proceed, we must assume, first, that supersymmetry is broken in a suitable way: scalar quarks and leptons, gluinos, etc., get mass of order the W mass, SU(2)×U(1) is broken, and so forth. We must also assume that the string coupling is weak. Otherwise, we can't compute anything. For example, if we wish to perform the standard Georgi, Quinn and Weinberg[36] calculation of the unification scale and $sin^2\theta_W$, we must assume that the unified coupling constant is weak.

More drastically, we have no guarantee that at strong coupling the light spectrum is what it appears to be at weak coupling.

With these assumptions, a great deal of effort has gone into developing a string phenomenology. Most work on the subject addresses the following issues:

A. *Proton decay:* First, one must make sure that the unbroken gauge symmetry does not allow gauge-boson induced proton decay. This already severely limits the possible low energy gauge groups.[37] But this is by no means enough. In supersymmetric theories, the problem of proton decay is quite severe since there are operators of low dimension which can contribute. Examples are dimension four operators, such as three quark operators, in the superpotential and dimension three operators which softly break supersymmetry, such as products of three scalar quarks. This is unlike the case of the standard model, where the lowest dimension operators (allowed by gauge symmetries) which violate baryon number posses dimension six. Some of the dangerous operators might be forbidden by discrete symmetries. For example, a symmetry which flipped the sign of all quark and lepton fields, but not some Higgs doublets, would forbid the dimension three and four operators. Dimension five operators could also be adequately suppressed by this symmetry, provided that any colored partners of the Higgs doublets were sufficiently heavy. We have seen above that such a splitting can naturally be achieved in string theory. However, to date no one has exhibited a manifold with an appropriate discrete symmetry. For an introduction to the problem of identifying discrete symmetries of Calabi-Yau spaces, I must refer you to Ref. 26. Some other ideas for suppressing proton decay will be discussed below.

B. *Georgi, Quinn, Weinberg Calculations:* One would like to find reasonable values, for the unification scale, unified coupling, and Weinberg angle. Weak coupling requires that the unification scale be of order M_p [8,32], and this an additional severe constraint. Among models with E_6 unification, in compactifications with four or more generations QCD is not even asymptotically free at high energies. Such models are only acceptable if there is a large intermediate scale ($M_I > 10^{11}$ GeV) at which many particles (essentially the extra $SU(5)$ 5 and $\bar{5}$) gain mass. In Ref. 37, a natural mechanism by which such masses might arise was suggested. Recall that, in global supersymmetry, the scalar potential is given by

$$V = \sum_i \left| \frac{\partial W}{\partial \phi_i} \right|^2 + \sum_a D_a^2 . \tag{25}$$

where W is the superpotential, i runs over all the chiral fields of the theory, and D^a is an auxiliary field associated with the a'th gauge generator,

$$D^a = \sum_i \phi^{i*} T^a \phi^i \,. \tag{26}$$

In the SU(5) decomposition of the 27, there are two singlets (Eq. 19), call them S_1 and S_2. It is easy to check that these appear at most linearly in the cubic terms in the superpotential, as a consequence of gauge invariance (e.g. extra $U(1)$'s which are preserved by Wilson lines). Thus if we set, for example, $S_1 = \bar{S}_1 = v$, where $\bar{S}_1$ is a scalar in the $\overline{27}$, all the terms in the potential vanish identically. This degeneracy will in general be lifted by dimension five terms in the superpotential, corresponding to dimension six terms in the potential. If these scalars also receive negative mass-squared of order M_W from supersymmetry breaking, they will have a potential of the form

$$V = -M_W^2 |\phi|^2 + \frac{|\phi|^6}{M^2} \,, \tag{27}$$

where M is some large mass associated with the string, such as M_p. They will thus obtain vev's of order $\sqrt{M_W\, M_p}$. These vev's, in turn, give masses to particles in the 5 and $\bar{5}$ of Eq. 19, and break some of the remaining gauge symmetry (e.g. extra unwanted U(1)'s). In Ref. 39, it was observed that the dimension five terms are sometimes forbidden by discrete symmetries, and the intermediate scale can in fact be larger. This may help with the problem of proton decay and also with obtaining the correct pattern of low energy gauge symmetry breaking.[38] This idea certainly deserves further exploration. Even with these assumptions, few models satisfy the renormalization group constraints.[37] In particular, it is difficult to use the mechanism of Refs. 26-27 to obtain light doublets while obtaining an acceptable value for $sin^2\theta_W$.

C. *Neutrino masses:* One needs to understand, in these theories, why neutrinos are so light, since, at least for vacua with E_6 unification, neutrinos may obtain Dirac masses. I believe the most promising suggestion in this regard is the possibility that higher dimension terms in the superpotential may give large Majorana masses to right-handed neutrinos, allowing a "see-saw" mechanism to operate.[39] Alternatively, orbifolds with SU(5) or O(10) unification might solve this problem.

D. *Other problems:* Of course, there are numerous other questions which must be investigated. One is the strong CP problem. There is always at least one axion in compactifications of string theory[3], but we may need more. Furthermore, even the existence of an axion does not yet guarantee a solution to the strong CP problem[44]. One additional candidate Peccei-Quinn symmetry has recently been shown to be badly broken.[25]. Another problem, stressed by Georgi[40], is that of strangeness-changing neutral currents. In supersymmetric theories, such processes are suppressed only if there is a high degree of degeneracy in the scalar quark mass matrix. How such a degeneracy would arise in string theory is by no means clear. Some of the issues discussed here are reviewed in more detail in reference 45.

SUPERSYMMETRY BREAKING, THE COSMOLOGICAL CONSTANT AND THE DILATON EFFECTIVE POTENTIAL

The cosmological constant is essentially the energy density of the ground state. In a free field theory, it is given by

$$E_0 = \frac{1}{2}\int d^3k \sum(-1)^F \sqrt{k^2+m_i^2} \tag{28}$$

where the sum is over all physical helicity states. This is, of course highly divergent, so we must introduce a cutoff, Λ. Even if Λ were as small as 100 GeV, this would give a result for E_0 10^{55} times larger than the present limit. (The universe is really amazingly flat.)

If nature were exactly supersymmetric, it would be easy to understand the vanishing of the cosmological constant: the boson and fermion contributions to Eq. 28 would simply cancel pairwise. This would continue to hold to all orders in perturbation theory in a weakly interacting theory, due to non-renormalization theorems. Once supersymmetry is broken, however, we would expect a result at least as large as M_W^4, and usually much larger ($E_0 > 10^{44}$ GeV^4).

In fact, in vacua of string theory in which supersymmetry is broken at tree level, one finds a cosmological constant already at one loop.[41]. These computations also yield a potential for the dilaton and the fields which describe the size of the internal space. This may be understood by simple scaling arguments. Start with the Lagrangian of Eq. 1. If we reduce to four dimensions, we still have a factor of ϕ^{-2} out front. The one loop vacuum energy then has no factors of ϕ. However, to obtain the standard form for the Einstein term (no factors of ϕ), we must rescale the metric

$$g_{\mu\nu} \to \phi^2 g_{\mu\nu}\ . \tag{29}$$

This puts a factor of ϕ^4 in front of the cosmological term,

$$\int d^4x \phi^4 \Lambda \sqrt{g}\ , \tag{30}$$

which is a potential for the dilation which goes to zero rapidly as $D \to \infty$, i.e. as we approach weak coupling. A similar scaling argument shows that the potential also goes to zero as $R \to \infty$, i.e. as we approach flat ten dimensions.

For compactifications which preserve N=1 supersymmetry, we expect that supersymmetry remains unbroken to all orders of perturbation theory and that no cosmological constant is generated. Supersymmetry had better be broken non-perturbatively, if string theory is to describe our world. One suggested source for this breaking is gluino condensation in the other E_8.[42] We will not review the mechanism here. Suffice it to say that this yields a dilation potential which goes to zero (for fixed radius) as $e^{\frac{-c}{\phi^2}}$, where c is a constant. Again, the system is driven to zero coupling.

In fact, these phenomena are quite general.[6] Any mechanism for supersymmetry breaking will turn off as the coupling goes to zero. The dilation effective potential thus always tends to zero as $D \to \infty$. One might hope that somehow supersymmetry is broken for any expectation value of D, but that the cosmological constant vanishes. But this possibility is ruled out, because it implies a massless scalar field whose couplings spoil the successes of Einstein's theory. Cosmological solutions are similarly ruled out.

The best one can hope is that, at strong coupling, there exists an isolated, four-dimensional ground state with broken supersymmetry and zero cosmological constant. I stress strong coupling in the sense that no weak coupling expansion is valid, not merely that the coupling is of order one. Any weak coupling analysis necessarily gives this unacceptable behavior for the dilaton potential.

What, then, of the phenomenology we discussed in the last section? It is certainly suspect, but for the moment it is the best we can do. We may hope that it will guide us to some correct picture. As for more "fundamental" work in string theory, we need some principle which might suggest how the cosmological constant could vanish after supersymmetry breaking.

Acknowledgements

I wish to thank the Weizmann Institute for its hospitality during the preparation of these lectures, and N. Seiberg for discussions of the physics contained here. This work was supported in part by DOE contract .

REFERENCES

1. For reviews, see: J.H. Schwarz, Phys. Rep. **89** (1982) 223; M.B. Green, Surveys in H.E. Physics **3** (1982) 127.

2. D.J. Gross, J.A. Harvey, E. Martinec, and R. Rohm, Phys. Rev. Lett. **54** (1985) 502, Nucl. Phys. **B25** (1985) 253 **B267** (1986) 75.

3. E. Witten, Phys. Lett. **149B** (1984) 351.

4. P. Candelas, G. Horowitz, A. Strominger, and E. Witten, Nucl. Phys. **B258** (1985) 75.

5. L. Dixon, J.A. Harvey, C. Vafa, and E. Witten, Nuclear Phys. **B261** (1985) 678; "Strings on Orbifolds II", Princeton Preprint (1986).

6. M. Dine and N. Seiberg, Phys. Lett. **162B** (1985) 299; and in **Unified String Theories** ed. by M. Green and D. Gross (World Scientific, 1986).

7. A.H. Chamseddine, Nucl. Phys. **B185** (1981) 403; E. Bergshoeff, M. de Roo, B. de Wit, and P. van Nieuwenhuysen, Nucl. Phys. **B195** (1982) 97; G.F. Chapline and N.S. Manton, Phys. Lett. **120B** (1983) 105.

8. M. Dine and N. Seiberg, Phys. Rev. Lett. **55** (1985) 366.

9. E. Witten, Phys. Lett. **155B** (1985) 151.

10. J.A. Shapiro, Phys. Rev. **D11** (1975) 2937; M. Ademollo et al., Nucl. Phys. **B94** (1975) 221.

11. C. Itzykson and J.B. Zuber, **Quantum Field Theory**, (McGraw Hill, 1985), p. 521.

12. This argument was developed in collaboration with N. Seiberg. Similar conclusions have been reached by E. Witten (private communication).

13. C.G. Callan, E.J. Martinec, M.J. Perry, and D. Friedan, Nucl. Phys. **B262** (1985) 593.

14. A. Sen, in **Unified String Theories**, edited by M. Green and D. Gross (World Scientific, 1986).

15. For a review, see D. Friedan, E. Martinec, and S. Shenker, "Conformal Invariance, Supersymmetry, and String Theory," Princeton University preprint (1986).

16. A.M. Polyakov, Phys. Lett. **103B**, 207 (1981).

17. D.H. Friedan, in **Les Houches Summer School**, 1982, edited by J.B. Zuber and R. Stora (North Holland, Amsterdam, 1984).

18. L. Alvarez-Gaume and P. Ginsparg, in **Anomalies, Geometry, Topology**, edited by W.A. Bardeen and A.R. White, (World Scientific (1985)).

19. S. Weinberg, **Gravitation and Cosmology** (Wiley (1972)).

20. C.M. Hull and E. Witten, Phys. Lett. **160B**, 398 (1985).

21. D.Z. Freedman and P.K. Townsend, L. Alvarez-Gaume and D.Z. Freedman, Phys. Rev. **D22** (1980) 846, Commun. Math. Phys. **80** (1981) 282.

22. D. Zanon, M. Grisaru, and A. Van de Ven, Brandeis University Preprint (1986), C.N. Pope, M.F. Sohnius, and K.S. Stelle, Imperial College Preprint, Imperial /TP/85-86/16 (1986); M.

23. C.N. Pope, M.F. Sohnius, and K.S. Stelle, Imperial College Preprint, Impe-

rial /TP/85-86/16 (1986); M.D. Freeman and C.N. Pope, Imperial College Preprint Imperial /TP/85-86 16 (1986).

24. E. Witten, unpublished; A. Sen and D. Nemeschansky, SLAC preprint (1986).
25. M. Dine, N. Seiberg, X.G. Wen, and E. Witten, "Non-perturbative Effects on the String World Sheet," Princeton University preprint (1986).
26. E. Witten, Nucl. Phys. **B258** (1985) 75.
27. J.D. Breit, B.A. Ovrut, and G. Segre, Phys. Lett. **158B** (1985) 33; A. Sen, Phys. Rev. Lett. **55** (1985) 33.
28. A. Strominger and E. Witten, Comm. Math. Phys. **101** (1985) 341; A. Strominger, Santa Barbara preprint (1985), and in Unified String Theories, edited by M. Green and D. Gross (World Scientific, 1986).
29. E. Witten, "New Issues in Manifolds of SU(3) Holonomy," Princeton preprint (1985).
30. E. Cremmer, S. Ferrara, L. Girardello, and A. Van Proeyen, Nucl. Phys. **B212** (1983) 413.
31. A. Strominger, Santa Barbara preprint (1986).
32. V. Kaplunovsky, Phys. Rev. Lett. **55** (1985) 1033.
33. E. Martinec, Princeton University preprint (1986).
34. C.S.Hamidi and C. Vafa, "Interactions on Orbifolds," Caltech preprint (1986), L. Dixon, D. Friedan, E. Martinec, and S. Shenker "Conformal Field Theory of Orbifolds," Chicago preprint (1986).
35. E. Witten, Nucl. Phys. **B188** (1981) 513; I. Affleck, M. Dine, and N. Seiberg, Nucl. Phys. **B256** (1985) 557.
36. H. Georgi,H. Quinn, and S. Weinberg, Phys. Rev. Lett. **33** (1974) 451.
37. M. Dine, V. Kaplunovsky, C. Nappi, M. Mangano, and N. Seiberg, Nucl. Phys. **B259** (1985) 549.
38. B.R. Green, K.H. Kirklin, P.J. Miron, and G.B. Ross, "A Superstring Inspired Standard Model" and "A Three Generation Superstring Model", Oxford preprints (1986).
39. F. del Aguila, G. Blair, M. Daniel, and G.G. Ross, CERN preprint (1985).
40. H. Georgi, Harvard University preprint (1986).
41. R. Rohm, Nucl. Phys. **B237** (1984) 553.
42. M. Dine, R. Rohm, N. Seiberg, and E. Witten, Phys. Lett. **156B** (1985) 55; J.-P. Derendinger, L.E. Ibanez and H.P. Nilles, Phys. Lett. **155B** (1985) 65.

43. A. Strominger, Santa-Barbara preprint (1985).

44. M. Dine and N. Seiberg, Nucl. Phys. **B273** (1986) 109.

45. N. Seiberg, to appear in the proceedings of the Third Jerusalem Winter School - Strings and Superstrings.

DISCUSSION

CHAIRMAN: M. Dine

Scientific Secretaries: M. Carter, F. Quevedo, P. Shotton and D. Rohrlich

– *P. Fisher:*

You seem to spend a lot of time and effort making sure the proton is stable. The experimental limit on the proton lifetime is $\sim 10^{34}$ years, Why are you concerned about making sure the proton is stable?

– *M. Dine:*

I was concerned about decays of the proton which would give a lifetime of the order of weak interaction lifetimes, $10^{-10} - 10^{-13}$ sec.

– *K. Meissner:*

Why do we not use bigger than $SU(3)$ holonomy groups to break E_8 for example directly to $SU(5)$ or $SO(10)$?

– *M. Dine:*

It may be possible to obtain unification groups other than E_6 like $SU(5)$ or $SO(10)$. As I mentioned this morning there have been attempts to build such models, however most of them turned out to be non–conformally invariant. Notable counter examples are orbifolds which is one of the reasons I suggest they should be studied.

– *S. Carlip:*

We heard from Prof. Gross that if the extra E_8 of the heterotic string is unbroken, the universe would be filled with massive glueballs which decay into gravitinos. This would presumably fill the universe with incoherent gravitational radiation. We have observational limits on such incoherent radiation in particular frequency ranges. Has anyone ever considered observable effects of string theory such as this?

– *M. Dine:*

What happens depends on the behaviour of the E_8. If it stays unbroken, it becomes strong very fast, and you have very massive states which have no symmetries to protect them, so they decay almost immediately in equilibrium. There does not seem to be any obvious observational effect. Another popular use for the extra E_8 is to use a gluino condensate to break SUSY, but this does not seem to work very well.

– *S. Carlip:*

What would the states of an unbroken E_8 decay into?

– *M. Dine:*

All sorts of junk – gravitons, gravitinos, dilatons, ordinary radiation etc.

– *M. Carter:*

You said that it is generally easier to do analyses on orbifolds. Can I explicitly calculate a Yukawa coupling on an orbifold as an integral of fermion, and boson wave functions without using topology?

– *M. Dine:*

Yes. But in practice that is not what you do. The problem is that an orbifold is a singular limit of a Calabi–Yau space and we cannot claim the interaction still comes from $\int d^{10} \times \overline{\Psi} A \Psi$. What you should really do is find the vertex operators for these states. This has recently been worked out by Vafa and Hamidi at Harvard and Dixon, Frieda, Martinec and Shenker. Orbifolds are interesting to study because the free field theory is tractable.

– *M. Carter:*

Does your analysis radically change if the manifold is not Ricci flat?

– *M. Dine:*

No. It is pretty much the same. People have considered these spaces as perturbations of the Ricci flat ones. To lowest order the β–function equation ensuring conformal invariance is

$$\begin{aligned} \beta_{IJ} &= R_{IJ} + \text{higher order terms} \\ &= 0 \end{aligned}$$

This equation is complicated to solve. To solve this you can develop a perturbation expansion. Actually doing this and writing the higher–order corrections is not easy. However, the non–renormalization theorem I mentioned this morning is still valid, since it applies to a procedure in which you correctly integrate out heavy fields by solving their equations of motion. It can be used to show that certain fields are massless, Yukawa couplings vanish and there are no tadpoles for massless paricles.

– *J. Quackenbush:*

Since you just assume SUSY breaking at low energies, how do you extrapolate from the unbroken to the broken theory?

– *M. Dine:*

I assume that the low energy theory should look like some of the supergravity theories that have been analysed in recent years, that is, that they look like N=1 supergravity with soft breaking terms. For example you can assume mass terms for the squarks and sleptons and introduce soft breaking terms. If you suppose this can be done you can set sensible limits and rule out some of the possible ground states of the theory. Of course, the mechanism of SUSY breaking is not known but the point is that even if we assume a suitable breaking mechanism, we can still rule out many models.

– *A. Shapere:*

You said that flavour–changing neutral currents require a degeneracy of quark mass levels. Can you explain the source of this constraint?

– *M. Dine:*

Recall how the G.I.M. mechanism works in $K^0 - \overline{K}^0$ mixing the diagram:

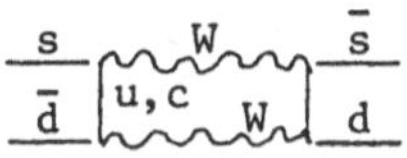

is of order $(M_W)^{-2}$. But when we sum over internal quarks, the diagrams cancel (to lowest order in the quark masses). so the process has an amplitude of order $(M_W)^{-4}$. Now consider the supersymmetric partner of this diagram, with squarks and winos.

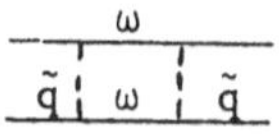

this gives an amplitude of order

$$\frac{1}{M_W^2}\left(\frac{\Delta M^2}{M_W^2}\right)$$

with

$$\Delta M^2 = (M_{\tilde{u}})^2 - (M_{\tilde{c}})^2$$

In the so–called standard supergravity model, to achieve $\Delta M^2 \simeq 0$, one postulates an approximate flavour symmetry between $\tilde{u}$ and $\tilde{c}$. There is no clear understanding of how this flavour symmetry might arise in string theory, which is certainly a very serious problem.

– *M. McGuigan:*

Could you explain why attempts at breaking supersymmetry through E_8' have failed?

– *M. Dine:*

Consider the super partner of the dilaton, let us call it χ. Its supersymmetry transformation law involves

$$\delta\chi = \lambda^a\lambda^a + \ldots$$

A gluino condensate $\langle\lambda^a\lambda^a\rangle$ thus gives $\delta\chi = \langle\delta^a\delta^a\rangle\epsilon$ and supersymmetry is broken. The problem is that the gluino condensate generates a vacuum energy $V_0 \sim \langle\lambda^a\lambda^a\rangle = (\Lambda_{E_8'})^6$ where $\Lambda_{E_8'}$ is the analogue of Λ_{QCD}. As with any grand unified theory we can calculate $\Lambda_{E_8'} \approx e^{-1/\pi g^2} M_{GUT}$. But as I showed before, $g^2 \sim <\phi>^2$. So V is an effective potential, which goes rapidly to zero as $D \to \infty$.

– *F. Quevedo:*

Can you mention the advantages and/or disadvantages of considering solutions with torsion i.e. with nonvanishing $H_{\mu\nu\rho}$?

– *M. Dine:*

These solutions have been studied; they are as interesting as any other. Phenomenologically they might be good for providing solutions with $SU(5)$ or $SO(10)$ symmetry groups. Also, it is now known that, even in Calabi–Yau compactifications a $B_{\mu\nu}$ potential is induced in σ–model perturbation theory.

– *Z. Bern:*

Would you be able to say something more about the differences between orbifolds and manifolds?

– *M. Dine:*

The answer to this question will appear in the written notes therefore it will not be repeated here.

– *M. Quiros:*

Flux breaking loops provide degenerate vacua. Do you know of any way of breaking this degeneracy?

– *M. Dine:*

The problem is ultimately connected with the problem of supersymmetry breaking for which, as I have said, I do not have a solution. If supersymmetry is broken, this degeneracy is probably lifted.

– *M. Quiros:*

Why did you claim that no axion–like fields in superstring theories can solve the strong CP problem?

– *M. Dine:*

I did not say that superstring theories cannot solve the strong CP problem. In superstring theories we typically have at least two strongly interacting gauge groups. The strongest one will give a large mass to one axion so we need at least two axions to solve the ordinary strong CP problem. One always gets one axion from $B_{\mu\nu}$ where μ and ν are indices in M^4. One hoped to get additional axions from B_{ij} where i and j are indices in M^6. However, it has recently been shown that world sheet instantons spoil this would be Peccei–Quinn symmetry, so some other solution must be found.

– *P. Vecsernyes:*

Is there any consequence from the phenomenological point of view if we are looking for such solutions of the string equations which are only logically product spaces?

– *M. Dine:*

I do not know.

– *G. Miele:*

I would like to know if there are any differences between using topological symmetry breaking methods to break E_6 in string theory, and adding at the Higgs representations an adjoint representations like the 78, in field theory.

– *M. Dine:*

In field theory I do not know a way to obtain massless doublets without colour triplet partners without fine tuning. Also this mechanism spoils the usual E_6 relations for Yukawa couplings, which is also good news.

A DISCUSSION OF THE SUPERWORLD

Moderator: A. Zichichi

Scientific Secretaries: P. Fisher, J. Quackenbush, S. Webb

Participants:

A. Ali
L. Alvarez–Gaumè
M. Dine
G. Ekspong
P. Fayet
S. Ferrara
S.L. Glashow
D. Gross
S. Raby
C. Rubbia
J. Schwinger
P. Van Nieuwenhuizen

– *A. Zichichi*

The Super World has attracted the interest of many, very many, physicists: especially the theoretical ones, to be more precise.

An enormous number of papers have been, and continue to be, published. The great trouble is that all these papers violate the Galilean principle (revived here in Erice by I.I. Rabi): which experiment should be implemented in order to prove if a theory is right or wrong?

Taking advantage of the very distinguished group of physicists present at the School, it seemed to me appropriate to have a discussion on the Super World.

Our activity is in a complex plane. On the imaginary axis, there are the theoretical speculations. On the real axis the experimental results.

The theoretical speculations are of great value if and only if, they have a non-zero projection, sooner or later, on the real axis.

For this discussion I would like to remind all contributors that the non-zero projection on the real axis is of crucial value for theoretical physics to be and remain alive.

– *A. Ali*

I want to put forward the view of a practitioner, someone who works very closely with experiment. From this point of view there are agonies in trying to understand what is going on. For me, the biggest problem is that there is no theory, I mean a theory with which you can calculate things.

In non–supersymmetric theories, we have the standard model. It is not perfect; it has many parameters and many problems, which have been discussed. But with it you can calculate things, its predictions can be compared with experiments, and you can draw conclusions. When there were alternatives proposed in the past, there always were some unique unmistakable prediction, and a theory stood or fell on that count. For example, there were attempts to replace the Higgs field with composite Higgs in models such as technicolour and hypercolour (there were many specific models created). It was soon shown that it is difficult to write a model that generates fermion masses which could also coexist with experimental limits on flavour–changing neutral currents. I would like to take that example more seriously. Low energy superstring models which have been discussed include many isosinglet quarks and isosinglet fermion fields. In general, the presence of these causes many problems with flavour–changing neutral currents similar to those which appeared in technicolour models. I am not sure whether low energy superstring models of this variety can survive these tests. The tests have not been made, however, as there is no consensus among the theorists as to what the low energy string theory is. It should make unmistakable predictions such as a difference between the $sin^2\theta_W$ value derived from the W and Z masses, and that measured from neutral currents, or it should predict a well defined particle spectrum with well defined couplings. If these predictions should compare with what we measure, wonderful. If not, then there is something wrong with the way in which we extrapolate down from high energies.

For me, it is essential that we should have a consensus from the superstring lobby on what the low energy superstring model is, because only those well defined models can be compared with experiments. Right now, even the general guidelines for an effective unique low energy theory are not available.

The present superstring theories are like letters of intent written by a lobby of theoretical physicists. They are very good in intent; but often what is said in the letter of intent and what is measured in the experiment are two very different things. The figure of merit of a theory is its predictive power which could be tested in an experiment in a laboratory.

– *S. Raby*

We want to understand how physics could be beyond the standard model. Hopefully pretty soon we will see some experimental evidence of something beyond the S–M, but before that evidence comes we must have some clues as to what direction to go in. The clues come from questions unanswered in the S–M. For example: quark and lepton masses and the mixing angles are parameters in the S–M; you don't know where they come from. The ratio of weak mass to the Planck mass (which is related to the theory of gravitation) also enters as a parameter and we have no idea where this small number comes from. Whenever you have a small number in physics you like to know where it comes from. For example, the small ratio of M_π/M_n is understood through dynamical symmetry breaking in strong interactions, which makes $M_\pi \neq 0$. We like to explain the ratio of the weak scale to the Planck scale in the same way (the gauge hierarchy problem). The many attempts made over the last few years to understand this are all equally valid: technicolour theories, composite fermions, and SUSY theories. Theoretically, composite and technicolour models lead to ugly models. SUSY probably provides the best answer to the gauge hierarchy problem.

– *P. Van Nieuwenhuizen*

I don't understand how any theorist can have a strong conviction about what he should do; I think everybody should be totally free to do what he finds interesting. The results will tell you what is viable and what is not. If you are interested in a field which is dead you don't get results. If you are interested in a theory and everybody says you are crazy that is all the more reason to do it. If you get good results it will show.

I don't think we should call the topic "superstrings"; we should call it "super–ideas". The point is not so much the particular model, each has its advantages; it is that we grope for new ideas. It may or may not be that fermion–boson symmetry is the thing. Everybody may have a conviction but it should not become a religion. Maybe some of us should not forget that. Some people will look for new structures. It is possible that through superstrings or supergravity we will discover new structures of spacetimes; nobody can predict that. I think everyone should be free and then results will always justify whether your hunch was right or not. It is very wrong from the beginning to tell someone to do this or that.

– *P. Fayet*

I agree more or less with what Ali was saying before. But there is another comment I would like to make. Ali was saying that superstrings do not make any predictions. It is a very nice theory, but it is only a mathematical theory, and I am not sure at this stage we can say it is a theory of particles. The situation is different if you think about supersymmetry and supergravity. SUSY makes well defined predictions: a lot of new particles. And while they haven't appeared yet they may appear in the future. It makes well defined predictions about the couplings of those particles, but unfortunately there are not so well defined predictions about their masses.

– *M. Dine*

People work on strings for various reasons. For me they have been fun to work on because they have pretentions to be truly unified theories and you can ask them almost all the questions you have always wanted to ask. You can ask about fermion masses, about generations, about the origin of gauge hierarchies and that makes them extremely interesting. You can ask them questions in a well defined, self-contained way, which has not been true before. String theories will not necessarily give the right answers and so there are serious questions about whether string theories can really describe nature. I also agree with Ali that these theories must ultimately make a low–energy statement, but I think it is quite shocking that a theory that is so complicated has, in a relatively short time, provided a framework in which you can address certain low–energy questions.

I would say they are fun to work on and probably the most interesting game in town. If I had to bet my life I would not bet on them holding ultimate truth, but they are fun in the meantime.

– *S. Ferrara*

My remarks will be related to what Fayet has said. If you are looking for finite supersymmetry field theories, without including gravity, there are many of them such as $N = 2$ and $N = 4$ Yang–Mill theories. The problems with finiteness only come when you want to include gravity and this is the only motivation for considering string theories. In that context the crucial part of having a finite string theory is SUSY because we know that non–SUSY string theories are not finite. So the links are gravity on the one hand and SUSY on the other. These are the two basic reasons why one is for this dramatic change from a local field theory to a field theory of extended objects. In that respect string theories are dynamical theories of supergravity; they are dynamical theories of gravity incorporating Bose–Fermi symmetry. In that respect I would like to see some signal that would distinguish string phenomenology from standard supergravity phenomenology. In particular we know that GUT's, which extended the standard model to the grand unification scale, made some predictions; for example proton decay. I do not see any analogy of proton decay for string theories, and I do not know of any prediction which is a crucial string effect and which cannot be predicted by supergravity theories. So, in the future, I would like to see something which can distinguish these theories by standard fields limits.

– *C. Rubbia*

I am afraid I am one of the few experimentalists here. In fact, I can see we are really getting fewer and fewer. I feel like an endangered species in the middle of this theoretical orgy. I am truly amazed. The theories are inventing particle after particle and now for every particle we have there is a particle we do not have, and of course we are supposed to find them. It is like living in a house where half the walls are missing and the floor only half finished. You should know how hard it is to discover a single particle in one lifetime and you come trying to lure us into discovering 24 particles with a new machine. It seems difficult to reconcile with the reality of scientific research. There is a very large separation between the way physics is being seen in the theoretical community and what really happens one flight below on the floor of the experiments. It may well be that for the book we have there will be another book with a twiddle on top. Some kind of realism has to come to the theory and you should not invent particles just to keep us busy. You promise us a bonanza of particles so that we go back to work rather than sitting and listening to you fellows. Still it is harder than you think to find new things.

– *G. Ekspong*

I would like to address the question of truth and beauty; truth being experiment, beauty being theory. We know that poetry may be beautiful but not necessarily true. One line of poetry is written on the podium, in front of us all $[(\delta + m)\psi = 0]$ which Dirac wrote in 1927. It is a beautiful line full of life and beauty. I think every theorist invent to do something similar, and when they see something beautiful they think it is also true which need not be the case. The line led to true predictions, such as the existence of positrons, but others were not fully observed by nature; that gave Schwinger some work to do. The problem is that the latest theories are so remote from experiment that they cannot even be tested. Therefore they don't play the same role as Dirac's equation. In the effort of trying to get to experiment there are many different roads which can be taken; Gross talked about 10^5 different roads. If one approach does not work, when you test it, try another. So it is difficult to test the basic ingredients of these theories.

The SUSY particles may not exist. One can always retract and say that the mass of them is higher than what has been reached so far. And then we must build something even bigger than the 100 TeV machine. It would be fine if theorists could make a prediction of masses; the mass problem is the essential problem of modern physics. If theorists could agree on a mass scale for various types of particles, we could really test their theories and make decisive experiments. I hope that this search for beauty does not drive theorists from experiments, because experiment has to be done at low energies, from one accelerator to the next and so on. Big jumps do not seem to be possible.

– *J. Schwinger*

Let me say that I consider my presence on this panel to be an excess of zeal by the management. It should be clear from the topics about which I have spoken at this meeting that I have not as yet concerned myself with the problems associated with strings. I should not have any opinion one way or the other. I do have opinions and they have been touched on by some of the other speakers on this panel. In particular, I second the plea that if we are to have super–anything it is super–ideas. Also, I do have the feeling that physics has been, until recently, an experimental subject, the theories also ought to have more contact with experiment. Or, to put it another way, theories should be more modest. They should try to make contact with what we know and to extend it a modest amount. Let us see if we can't avoid theories that attempt to be theories of everything and thereby announce that we have in our hands the theory of the whole Universe. If the answers were already in sight I would regard that as a very depressing future for you all.

– *S.L. Glashow*

It is a very sad time when we tell our students that it is all done or about to be done by the people in Princeton. I have to say I agree with my teacher, the previous speaker (J. Schwinger). Harvard is about to celebrate its 350th anniversary as a school. It began as a school of theology and it appears it may end as a school of theology. I have the greatest respect for string theory. It is a wonderful theory; it is a beautiful theory; it is an imaginative theory; as many of the previous speakers have said, it may or may not have anything to do with the physical world. It is for our experimental colleagues and for you to make the decision. Meanwhile, as it has been said, it is one of the only exciting games in town, which is why it attracts so much interest. Good, let it attract that interest; we shall see what experiment shall say.

– *D. Gross*

The last speaker expressed a feeling of frustration, a feeling that I share. We are all frustrated, these are difficult times. But I don't think that this frustration should cause physicists to resort to calling other physicists theologians; nothing is more insulting.

Also you have heard our experimental friends express their healthy suspicion of theorists. Those of you who are too young to remember the days in which we did not possess a theory which could account for all current experimental facts should be told that before the standard model the experimentalists, for good reasons, were infinitely more suspicious of the theorists. It is only recently that they have become slightly worried, because we have not only managed to produce explanations of some of the phenomena that they have discovered but have also managed to make predictions which they have confirmed. So they are a bit worried that we might be right, and future experiments will only discover things that we have predicted. Their scepticism is, of course, well founded. There is no question that they should go ahead and justify their future machines, not on the basis of supersymmetry or superstrings, but rather as tools for exploration, independent of prevailing theoretical ideas.

Similarly, my theoretical friends who would like to see directions other than string theory explored should do so. Students entering the field should certainly not work on string theory simply because it is the only game in town. It is very important, for both theorists and experimentalists to keep an open mind, develop a whole series of alternatives and ideas, and not have a situation where everyone is doing the same thing.

The problem we face is that there is a dearth of new experimental discoveries or of new theoretical ideas. This is summarized by a list that Luis Alvarez-Gaumé and I came up with last night when we were thinking about the alternatives to strings. We came up with two lists of alternatives-one for particle "phenomenology" and one for formal particle theory.

One of the important tasks of theorists is to accompany our experimental friends down the road of discovery; walk hand in hand with them, trying to figure out new phenomena and suggesting new things that they might explore. This is called phenomenology and, in periods of active experimental research, is the most important role for theorists. Today it is a difficult role since there aren't many new phenomena being discovered, and those that are have a very short mean life. We therefore came up with a list of recent discoveries that you could work on (instead of doing string theory). You might explore the Zeta 1.3 Higgs meson, neutrino oscillations, proton decay, monojets, magnetic monopoles, 17 Kev neutrino masses, Cygnus X-3, the 1.73 e^+e^- resonance, the fifth force or SDI.

The other side of particle theory are the speculative attempts to extrapolate beyond what is now observed. These attempts are only rarely successful and they often attract too many people. If you are interested in working along these lines and not doing string theory you might work on: mooses, extended technicolour, constructive field theory, non-standard quantum mechanics, measurement theory or SDI.

These alternatives to string theory are not very attractive. This is the reality that is behind some of the antagonistic feeling towards string theory. String theory is, admittedly, an extremely ambitious and chancy attempt to extrapolate very far from what we know, to a "theory of everything". This description of the goal of string theory has raised the hackles of some since it sounds like an arrogant and unlikely goal. I agree that string theory is not going to be a final theory of everything. Normally, at each stage in physics we ask a certain number of questions, which are formulated in terms of what we already know. As we learn more we begin to ask new questions. The so called "theory of everything" that we are now talking about pretends to address all of the questions which we are now able to pose. If the theory turns out to be correct then I am sure that there will be a whole set of new questions to be asked, which the theory cannot answer. So, don't worry, there will always be more to do.

– *L. Alvarez-Gaumè*

As one of the younger members of this panel, I would like to say that there are two aspects of doing theoretical physics that I always look for: it should be fun, and it should be challenging. I would not advocate that all the community should be doing superstrings, but there are some outstanding conceptual developments that are perhaps within reach of superstring theory. One such development is that superstring theory may offer a theory that can solve, or at least help us understand, a consistent quantum theory of gravity. Understanding the quantum structure of space–time and to extend the standard $SU(3) \times SU(2) \times U(1)$ model to the Planck scale has been a problem in theoretical high energy physics, and it has fostered many interesting ideas to accomplish such big synthesis. In my opinion, it seems that the only theory we have which might work to infinite energies is superstring theory and the conceptual revolution involved alone makes it worth pursuing. But again I would not advocate everyone in the audience or the world to do superstring theory. There are other aspects of theoretical physics worth pursuing. However, what motivates me to do superstring theory is that certain ideas and question which previously could not even be formulated in some physical or mathematical framework can now be phrased in terms of theories which might work to infinite energies. This is something which did not exist ten years ago.

– *G. Snow*

To me, string theory seems very exciting. The question in my mind is: suppose five years go by and there is still no concrete evidence of superparticles. Then what?

– *L. Alvarez Gaumè*

I could answer you by a different question, as the Japanese usually do. There is a very good example from Ed. Witten: it would have been very hard to think in 1930 that one would be able to see gravitational lenses, or gravitational radiation, or even neutron stars. It is not clear to me that the evidence for S–particles, SUSY, SST's is going to be in the most straightforward application of the ideas we have been following over the last ten years.

– *S.L. Glashow*

I am familiar with that example from your prophet. He was talking about 1930. Let us remember that in 1932 the positron and the neutron were discovered so there were a few other things happening aside from waiting around for gravitational lenses. That is not so obviously true today.

– *M. Carter*

We students are often encouraged to work on various things and to be very imaginative, but it seems that if you are not very interested in strings then your job possibilities are severely limited.

– *D. Gross*

Young people are often driven to work on fads for the wrong reasons, and employment might be one of them. As one who comes from Princeton, which is certainly a stringy place, I will tell you that the two postdoctoral offers we made this year were made, on purpose, to people who had nothing to do with strings and who, at least before they came to Princeton, knew little about the subject.

– *S. Giddings*

I have a question for Shelly Glashow. Everyone has seen your recent letter in Physics Today which was mildly critical of everyone doing strings. As a young theorist, what would you suggest as alternatives?

– *S. Glashow*

The article was not intended to be critical of SST; it was written jointly with a well known SST physicist. We expressed our fear that the theory might be a correct theory, but we would have no way of knowing that it is correct.

Concerning employment, the most important comment was made by David Gross; don't pick your field of interest based on where the employment opportunities lie, unless you want to do SDI. It is a tough time; many of the students at Harvard have chosen to work in astrophysics or geophysics or cosmology. Particularly cosmology seems to be moving rapidly at the moment, full of challange. Plenty of us are moving more and more toward that field, even though we don't know very much about it. As far as doing SST goes, it is a risky game. If you really have the stomach for it, do it, by all means. If you have doubts it ain't for you.

– *Wormser*

I would like to know about the proton decay experiments. They seem to me to have settled theoretic predictions that the proton lifetime would be approximately

10^{32} years. Was it really fair to give such a definite prediction? It seems it was very precise; just in the window where the experiment could be done.

– *S. Raby*

SUSY $SU(5)$ makes some pseudo predictions. There are no definite predictions coming out of $SU(5)$, the reason being that any prediction depends on a certain set of unmeasured quantities. For example, it depends on low energy SUSY spectrum, the mass of these particles and their couplings and the mass of a colour–triplet Higgs Fermion whose couplings are not related to gauge coupling constants. In order to know what these couplings are, we must know the Fermion mass spectrum and the Yukawa couplings. It is also related to the mass of this object, which has not been observed. So SUSY does not make any hard prediction for the proton decay rate. What they can do is make predictions for branching ratios for proton decay. The standard $SU(5)$ GUT made stronger predictions because the mass of the particle which mediated proton decay was a gauge particle whose couplings and mass you knew to a high accuracy.

– *C. Rubbia*

I would like to make a brief comment about proton decay because it seems to be an interesting example of how you can pollute an exquisite field of experimental science with theoretical prejudices. Of course some people think proton decay was invented by theorists. Proton decay is one of those things we had been looking for for many years; even before theory found it fashionable. I hope it will be looked at after theory has decided it is no longer interesting. In fact, I think what we need are some very good proton decay experiments.

Experimentalists should not be too influenced by whatever "brand X" theory tells you should do. A few years ago, we felt that proton decay was an easy way of proving $SU(5)$. We found that $SU(5)$ was not quite what we want. Now, proton decay seems to be dead. I feel that proton decay is fundamental, like β–decay. We must find an experimental method which allows us to find such a process cleanly. It is exquisitely an experimental issue and I hope that this research is continued independent of theoretical bias.

– *P. Fisher*

I am an experimentalist, so I am a spectator of the superworld, and I have a lot of trouble with the complicated language and more than four space–time dimensions. I would like to ask anyone on the panel if they can tell me some way, besides coming to superworld and talking to my theoretical colleagues, how to think about these things in a non–technical but useful way, if that is possible.

– *S. Glashow*

I like that question because it is a central question to my own thinking, because I think I am a frustrated experimentalist. I have not been able to find any answer to it whatsoever. Sometimes our SST people consider the low energy implications of SUSY but those that are honest basically say there are not unambiguous predictions of low energy SUSY. It is a real problem and it is the root of my aversion to string physics. Maybe a time will come when you and they can talk to each other on equal terms. But that time has not yet come and it is the fervent hope of the experimenters and of my colleagues who do SST physics that the connection that you and I desire will be made. But that time is not quite here.

– *F. Quevedo*

Some say that a criticism of superstrings is that they have not made feasible predictions to distinguish them from normal SUSY in 4D. Can we see that as encouragement that there needs to be more work done on the theory? For example, in 4D super gravity, there were no experiments needed to convince people that it was wrong.

– *D. Gross*

The last time that I was in a debate like this in Erice was in 1975, in a similar debate about the asymptotic freedom and QCD. At that time there was still much scepticism about the standard model, and many people were working on totally different approaches. But in reality it was all over, the main ideas behind the theory were laid down. A lot of work remained in applying and testing the theory, which goes on even today, but it was conceptually complete. This is certainly not the case today. There remains in string theory an enormous number of very deep mysteries. String theory is less of a theory, at the moment, then a direction of research or a framework for further exploration. It is not at all clear how much of the theory we now are aware of. It might be 10% or it might be 50%. For an experimentalist, who would like to be told how to test the theory, this situation is very frustrating, for a theorist who wants to make new discoveries it can be very exciting.

– *M. McGuigan*

I would like to make a comment, although the experimentalists are shut out now, there are other colleagues who have been shut out for a long time, and who are getting very interested in physics now and those are the mathematicians. They made a great contribution to physics in the last century and it would be nice to have them come back and work on physics again.

– *A. Zichichi*

These speculations are attractive. It does not mean that we experimentalists should do what theorists say and if they don't say anything, we should stop doing experiments.

– *M. Carter*

I always ask and I never get a nice answer. Why aren't people worried about the 500 or 1000 massless particles (in SUSY)? Why is that nothing to worry about?

– *L. Alvarez Gaumè*

I think we don't understand very well how the symmetry breaks in string theories. For example, although 3 is not like 496, but definitely the W^+, W^- and Z^0 are supposed to be massive and before spontaneous symmetry breaking and the Higgs mechanism were discovered no one really understood how to get the masses in a consistent way. Although I share your concern about what the natural mechanisms to produce masses and break symmetries in the theories are, I think it is a little bit premature to know what a consistent answer to the problem.

– *M. Dine*

Nobody knows how string theory works or if it works. Our experience is that particles that are not protected by symmetries acquire mass. In particular, most of the particles that appear in compactification of super string theories are not protected by gauge or global symmetries, apart from SUSY. The million dollar question is whether or not SUSY can break in some sensible way. If it could, that problem would go away. That problem looks minor compared to other problems the theory has to face.

– *F. Lizzi*

A comment on what Carter said. As a young person, I don't feel you work on what gives you a job, but you work on what you feel what you can do. This is a balance between how interesting you find the problem and how difficult and challenging the problem is.

– *A. Zichichi*

We all like super strings, the super world. We are fascinated by the new ideas going around and we would like to check them. To the students: could someone propose the crucial experiment to check what happens at the Planck mass?

– *P. Fisher*

I propose a different question. The question should be: "what is the next step in the development of the theory to work on?" In seven words or less.

– *S. Glashow*

In seven words? "My God, why has't thou foresaken me?"

– *P. Van Nieuwenhuizen*

I think people would like to know more serious alternatives, to SDI, for what theorists could do. Shelly mentioned cosmology. I would like to mention some alternatives. I know a lot of young theorists who are going to do radiative corrections for experiments. I did it long ago. It is not, at the beginning, very challenging work, but a lot of theorists will be needed to do it in the future. Another example is the work being done by 't Hooft; he thinks about the structure of space–time and black holes. And in Stony Brook, there is George Sterman who works on jets in experiments and he does computations. Some people claim that it is field, but there are a lot of unsolved problems and he seems to do good work. So I think that there are a lot of alternatives other than cosmology, and I invite my colleagues to contribute to this list.

– *C. Rubbia*

I would like to add one small element to your list. The old predictions on the experiments are good to within a factor of two. That factor of two has to be explained. Give us calculations which predict the fundamental strong interaction phenomena to within 5% and then let us test it and we will know whether there is a theory there or not.

– *J. Quackenbush*

The problem I have with QCD is that it can't predict to within 5%; there are a lot of problems. I wonder what will happen if string theory hits that plateau where all the easy problems are solved. Then what do we do? What is the next step?

– *D. Rohrlich*

If I think of the scientific revolutions of the thirties and before, they strike me as being very rich in intellectual content. When I think of the present developments in physics, I don't see the same thing, although they are mathematically very sophisticated and elegant. This brings to mind Weinberg's comment that physics is getting more and more meaningless.

– *D. Gross*

"Meaning" is not a straightforward thing, like the taste for good wine it must be acquired. If string theories are to work, it is clear that new concepts will be required. You say that the theory "is mathematically sophisticated, but where are the new concepts?" Well, some have already appeared, albeit in a primitive form, but more are on the way. Indeed, one of the motivations for studying string theory, for many of us, is the feeling that if we are to finally understand quantum gravity, and to construct a unified theory of gravity and matter, new concepts will be involved. Most likely these will be related to a fundamental length scale. The previous revolutions in physics that you referred to involved physics at velocities near the velocity of light and actions close to the Planck constant of action. The only fundamental unit left, in addition to c and h, is a fundamental unit of length and the obvious candidate is the characteristic length scale of gravity, the Planck length. Are there new concepts that will emerge at the Planck length? My feeling is that there will be. The problem is that we are many more orders of magnitude larger than this fundamental length then our predecessors were larger than theirs. So it will be more difficult to discover these new concepts. We hope that they will emerge as we push our understanding of strings to shorter distances.

– *S. Hands*

As a lattice theorist, it seems that string theories are lacking nonperturbative dynamical predictions. Our experience from lattice calculations is that these numbers are not easy to find. From where do the people proposing string theories feel that they will get nonperturbative numbers?

– *D. Gross*

Let us hope that it will not turn out to be as difficult as in QCD. As I explained in my lectures we must have nonperturbative input. However, there is one important difference that might make matters simpler than QCD. In the case of QCD the theory is simple in the ultraviolet and complicated in the infrared. In string theory we live in the infrared where the theory looks simple, or at least familiar, and we have no idea what it looks like in the ultraviolet. If there is any merit at all to string theory then the large-scale physics that we observe will require only small corrections arising from nonperturbative phenomena. Thus, for example, there need not be anything as complicated as confinement in string theory. The particles we hope to get out of the theory are not bound states in the way that hadrons are bound states of quarks and gluons. All we need are nonperturbative effects that softly break some of the unnecessary symmetries of the theory. We can hope that this task is simpler than it is to construct confined bound states in QCD.

– *S. Raby*

For many years people have been getting N=8 supergravity out of strings as a low energy theory. Since it does not have the right gauge content, people were claiming that these states were the preons and that the physical states would emerge in nonperturbative mechanisms. The argument is good even if we see the emergence of gauge fields that we might like. You don't know whether the asymptotic states of the theory are given nonperturbatively, or whether the states we like are the asymptotic states of the theory.

– *S. Ferrara*

I would like to ask the string theorists what they mean by nonperturbative supersymmetry breaking. Can you give an example?

– *M. Dine*

There are certain examples in field theory of calculable nonperturbative supersymmetry breaking. Whether an analogous example occurs in string field theory is certainly an open question. What makes QCD difficult is that it doesn't really have a parameter, so there is no controllable approximation that one can do. It is often said that string theory doesn't have any parameters, but as I think David has explained, string theory has several types of parameters and expansions. There is certainly a sense in which there is a weak coupling expansion in which you can exhibit the spectrum. That is, in fact, the appeal of string theories; what we know about it comes from studying that expansion. Whether the true ground state of the theory appears in that expansion remains to be seen.

– *C. Rubbia*

Last year I was told that the next step in string theory was to calculate lepton masses and other simple things. What are the physical quantities we expect to see emerging from the theory and how do we go about calculating them?

– *M. Dine*

There was enormous optimism after the discovery of Schwarz and Green that progress would be very rapid and all the questions one would like to ask would have received an answer. It has turned out that the hard problems which have arisen have not been easily solved.

– *C. Rubbia*

You have not answered my question. We already have experimental information about things which you are striving to predict. Can you predict them, and if so, with what accuracy?

– *M. Dine*

Ideally, you want to calculate everything, such as the quark and lepton mass matrices, the spectrum of quarks and leptons, the number of generations, and the values of the gauge couplings.

– *C. Rubbia*

Do you know how?

– *M. Dine*

At the moment, no. But it is not inconceivable that at some stage you might be able to.

– *A. Zichichi*

I would like to thank all the members of this panel. I encourage the theorists to be optimistic; we want them to be ambitious. The quantity of new ideas which have been produced is impressive, and we don't want them to be demoralized. We experimentalists must go on and build big machines to provide answers for the theorists.

Julian, would you like to make a closing statement?

– *J. Schwinger*

The future lies ahead.

THE END OF THE SUPERWORLD

Sheldon L. Glashow

Lyman Laboratory of Physics
Harvard University
Cambridge, Massachusetts 02138, U.S.A.

Our discipline of elementary particle physics confronts the three great pincers of the Big Squeeze: we suffer from a lack of adequate funding, a theory that works too well, and a bit too much religious zeal.

The standard model is more than a model. It is a theory. In its context, it offers a complete, correct, and consistent description of all *observed* particle phenomena. The trick work is 'context'. It doesn't solve the mystery of superfluous replication, the appearance of at least three isomorphic fermion families. It does not constrain the fermion masses. It is a theory with about 17 arbitrary and adjustable parameters. Surely the ultimate theory has fewer. Yet, the standard theory works. There is little reason to doubt that the rich spectroscopy of hadrons detailed in our 'wallet cards' is explained - at least, in principle - by quantum chromodynamics. At the moment, the explanation is generally qualitative. Future developments in computer-assisted lattice field theory is expected, someday, to provide a more quantitative description. However, there is every reason to believe that QCD is correct. There are no hadrons that defy understanding in terms of quarks. While some data may seem puzzling, there is none that flatly contradicts the theory. From our reductionist viewpoint, hadron spectroscopy has joined its nuclear and atomic predecessors as a mature discipline.

Much the same may be said about the electroweak theory. Its simplest realization seems to work perfectly. It has passed all of its tests. Only the Higgs boson remains to be detected. There has been a wide-ranging search for departures from the standard theory of weak interactions, but the results have been disappointingly and consistently negative. Why disappointingly? Why not triumphantly? Because we ardently desire a better theory, one that *will* tell us why there are muons and why they weigh what they do. If we are lucky, there are some hints about the better theory that can be found by experimenters. The hints will appear as failures of the standard theory, except that nobody so far has found a failure. Here are a few of many examples.

The minimal standard theory assigns zero mass to the neutrinos. They cannot have Dirac masses since there are no right-handed neutrinos in the standard model. They cannot have Majorana masses since there is no Higgs triplet in the standard model. And, there is absolutely no compelling evidence that this simple prediction is untrue. Accelerator or reactor evidence for neutrino oscillations is often put forward, and is just as often demolished by subsequent and more precise measurements. Simpson's heavy neutrino component seems to have bit the dust. The Swiss tritium experiment contradicts the Russian assertion of finite electron-neutrino mass. The solar neutrino puzzle, in all likelihood, simply shows that our modelling of stellar structure is not so good as some astrophysicists believe. And finally, the evidence for no-neutrino double beta decay has evaporated. As of now, the most plausible and simplest hypothesis is that neutrinos are massless. Just like the theory says.

The standard theory puts all fermions into canonical 15-dimensional anomaly-free chiral representations of the gauge group. More elaborate representations could have shown up in nature. But, there is no evidence for quixes or quaits, or for SU(2) triplets of fermions, or for doubly charged leptons, or for fractionally charged matter. Mysteriously, nature seems to make use of only the simplest of possible fermion families. More mysteriously, nature does it thrice. Or maybe more. One thing we should soon learn is whether or not there are more fermion families of the same ilk. The neutrino-counting experiments at SLC and/or LEP will be an essential test of the minimal (3-family) standard theory.

We may consider fiddling with the gauge group as well as with the fermion representations. The left-right symmetry gang has not given up. They are waiting patiently for evidence of a second and heavier W particle revealing another SU(2). Some superstring phenomenologists (forgive the oxymoron) believe in a second Z corresponding to another U(1). Yet, there is not a shred of experimental evidence justifying either enlargement of the gauge group. Once again, the simplest standard theory is the best one.

Various theorists, with or without tongue in cheek, have often argued that there may, should, or must be a host of new particles and new forces lurking just a bit further down the high energy frontier, from a few hundred GeV to a few TeV. All these wonders are just waiting to be discovered by the next accelerator to be built. Perhaps this is true. After all, we are just beginning to explore the physics at these energies. Perhaps the W and the Z will be accompanied by hosts of their unexpected and uninvited cousins. However, these marvelous surprises are severely constrained by the quantitative successes of the standard theory at lower energies.

Hypothetical new particles often can induce GIM forbidden decay modes or an anomalously large neutral kaon mass splitting. Yet, there is no evidence for any such anomaly. Careful searches for a vast variety of forbidden decays have confirmed just that: they are forbidden. There is no reason to suspect that $\mu \to 3e$ is lurking just around the corner. The most ingenious experimenters have not tossed a single pebble into the vast sea of placid tranquillity that the standard theory provides. The success of the theory is overwhelming and depressing. The last giant step was the discovery of the tau lepton, more than a decade ago. What, if anything, will be the next great experimental surprise? Do there remain any pieces to the puzzle that won't fit into our pat picture?

Maybe the next great discoveries are simply too costly for society to choose to afford. The pursuit of high-energy physics costs big money and involves incredibly complex instrumentation. LEP is big enough around to surround a fair-sized city, and will cost almost a billion dollars. The proposed (but still unapproved and unfunded) Superconducting Super Collider would cost three billion dollars to build, and the Eloisatron is a tera-lira toy. This is just the up-front money. Divide the cost of a machine by the construction period to obtain a crude guess of the operating expenses. Add a few hundred million dollars for detectors. Data analysis will challenge the world's largest and most expensive supercomputers. Sooner or later, someone will ask us at what level elementary particle physics *should* be supported. The Kendrew committee has not only asked the question, but has answered it: CERN cuts or Britain quits! The U.S. high-energy budget is under similar pressures. The tevatron collider and SLC may get finished, but if and when they do there may not be any money left to run them.

The budget squeeze is not something new. Both American and European funding for particle physics have been dropping rapidly during the past decade. The time will never come again when any accelerator that can be built will be built. We must become more selective and more cooperative as we grow leaner and hungrier. There has been very little real progress in recent years at the highest energy frontier. The CERN collider revealed the things we all knew were there, but little else. (True, there had been some smoke at one time, but it turned out to be a false alarm.) There has been no substantive competition from America since the CERN collider was first activated in '82. Years will pass before the Tevatron collider can achieve and surpass the accumulated luminosity of CERN. Things move with monumental slowness. Meanwhile promising and exciting fixed target experiments are suspended or deferred for lack of funds. Who knows how many wonderful discoveries have been overlooked? Graduate students enter Harvard, earn their degrees, and become professors while the field remains absolutely static. It has never been like this since the time of Columbus.

Clearly, a lot more could be done with a lot more money. We could build at least one (and one is enough) of the proposed super colliders, SSC, LHC, or Eloisatron. Moreover, we might even hope to see new physics coming from these machines during our professional lifetimes.

The third part of the threat to particle physics has to do with the superstring and the philosophy it embodies. Since the time of Galileo, Boyle, and Newton, progress in physics has followed the upwards path, from the phenomena towards a rational explanation of theory. The downwards path hadn't worked very well. However elegant it would have been had planets moved in circles, they simply don't. The revolution of special relativity emerged from observation and not from the power of positive thinking. Without the careful experimental work that led Maxwell to his equations, no philospher could have discovered the real nature of space and time. Quantum mechanics, too, crept out of an experimental morass which was irreconcilable with classical theory. It was not an abstract mathematical creation.

Nonetheless, many of the brightest young theorists are turning once again to the downwards path. True enough, superstring theorists have stumbled upon what may be the first consistent quantum theory of gravity. True enough, the superstring theory may be capable of encompassing the standard theory, although this has not yet been shown. Mostly, however, it is a long way down from the Planck mass to the observable world. It is a lonely path, along which further experimental data is neither necessary nor particularly desired. The superstring theory is so ambitious that it can only be totally right, or totally wrong. The only problem is that the mathematics is so new and so difficult that we won't know which for decades to come. Meanwhile, new experimental data, while perhaps a subject of idle curiosity, is largely a matter of indifference. The superstring theorists are doing their own thing, and it has little to do with anything else in physics. Superstring theorists are separating themselves from their discipline, just as their little six-dimensional ball is separated from the empirical realities of mortals.

While I often pick on my superstrung colleagues, I sympathize with and even envy them. They are faced with the real challenge of figuring out just what it is they are doing. Perhaps, one day they will even have a theory that says things that could involve the potential for empirical verification. Things are moving on the downwards path, while the upwards path is, for the moment, blocked. Our theory works too well, and we are too poor to seek further experimental clues. At least there are no tolls on the downward path. How things have changed!

The 1960's was a time of tremendous experimental discovery. Large bubble chambers were deployed at fixed-target proton machines, and dozens of new hadrons were discovered. Some explained the nucleon form factors, some lay on Regge trajectories, but all of them fit into representations of flavor SU(3). How could there be so many 'elementary' particles? Advocates of nuclear democracy and maximal analyticity argued that they were all equally elementary, and hence not very elementary at all. Others held out for another layer of the onion, and argued for the existence of more elementary constituents like quarks. Late in the 60's, quarks were 'seen' about as clearly as they could be in the deep inelastic electron scattering data at SLAC. Meanwhile, the violation of CP conservation and the existence of cosmic black body radiation were serendipitously discovered by scientists from New Jersey.

The next decade was no less exciting. Soon after 't Hooft's epochal work which made such experiments fashionable, the Europeans won the trans-Atlantic competition for the discovery of neutral currents. As curious data from SPEAR and the ill-fated CEA accumulated, Burt Richter convinced himself that the electron is part of the time a hadron. Then, with the twin discoveries of November 1974, the standard theory began to achieve popularity. The J/psi particle presented us with a new laboratory in which the quark model, the idea of charm, asymptotic freedom, and the newly invented doctrine of quantum-chromodynamics could all be tested, and would all be confirmed. Not only were charmed particles discovered, but nature surprised us with her unexpected third fermion family. It began with Martin Perl's tau lepton and continued with Lederman's upsilon particle. The top quark remains to be found, one of the last of the standard theory's missing links. In 1978, the last obstacle to the electroweak theory was removed. Atomic physicists had been managing to screw up the detection of the parity violating interaction between electrons and nuclei. In a beautiful electron scattering experiment at SLAC, the prediction of the standard theory was confirmed at last, and the way was cleared for the electroweak Nobel Prize.

What do the 80's have to say for themselves? Gluon jets were seen at DESY, and they are exactly as they were expected to be. Similarly for the W and the Z. There have, however, been no surprises at all in the elementary-particle physics game. More precisely, all of the surprises of the 80's (that is, things that don't square with the standard theory: like the zeta particle, the CERN monojets, the Darmstadt 'particle', the 17-keV neutrino, etc.) have been retracted, unconfirmed, or contradicted. Of course, there have been some interesting cosmological developments. Why are all of our neighbor galaxies within dozens of megaparsecs sharing in the Milky Way's peculiar motion? Why does the distribution of galaxies in the universe resemble the suds in a kitchen sink? Why does Cygnus X-3 act as a source of multi-TeV photons? How come nobody can detect any interesting inhomogeneity in the cosmic relic radiation?

Of course, the 80's are not quite over. SLC and the Tevatron Collider may still produce something unexpected. Maybe axions will be observed. Maybe the dark matter of the universe will be detected on Earth. Hopefully, the 90's will see the deployment of the Eloisatron or the SSC. There will be amazing and totally unexpected discoveries. There had better be, lest tomorrow's theorists are led to the 10^{68} GeV Superdrum, with today's superstring as its low-energy phenomenological limit.

In the remainder of my talk, I would like to describe some of the interesting things to think about that have nothing to do with the superstring: The Solar Neutrino Problem, Fischbach's Fifth Force, The Hubble Bubbles, and the possibility of An Electrically Charged Universe.

For over fifteen years, Ray Davis and his colleagues at Brookhaven have been observing the high-energy fringe of the solar neutrino spectrum, those neutrinos coming from the boron side chain. They see too few by a factor of three or so compared to the prediction of the 'standard solar model'. Either neutrinos are doing something funny on their way from Sun to Earth, or our model of solar structure is faulty. More particularly:

I. Perhaps neutrinos decay on their voyage, as Bahcall once suggested. There is no known mechanism for such a process, but one can be devised if necessary. Neutrino decay would selectively deplete the flux of *lower* energy solar neutrinos on Earth by the factor $\exp(-E'/E)$, where E is the neutrino energy, and $(E'/5\text{ MeV}) = (M/1\text{ eV})(10^{-4}\text{s}/T)$ where M is the neutrino mass and T is its lifetime.

II. Perhaps neutrinos undergo vacuum oscillations on their way from Sun to Earth. Maximal time-averaged oscillations among three species could result in a suppression by a factor of three, but it is hard to see why the oscillations should be maximal. If the oscillation parameter is chosen 'just so', then the typical electron neutrino has just enough time to transform itself into another species. Calculations show that such a choice can produce suppression in the Chlorine experiment by at least a factor of three over an octave of neutrino mass differences, with maximal

2-species mixing. As Lawrence Krauss and I note, 'just-so' neutrino oscillations lead to a measurable annual modulation of the observed Chlorine capture rate.

III. Perhaps neutrinos are subject to enhanced oscillation within the solar interior. This notion, first considered by Wolfenstein, is the subject of considerable recent attention. This effect should selectively suppress the *higher* energy neutrinos. If this is the resolution to the puzzle, essentially all of the solar neutrinos of greater than 5 MeV are expected to have become muon neutrinos.

IV. Maybe the assumptions implicit in solar modelling are false. In this case, the boron process is suppressed because the central solar temperature is a bit less than we thought. In this case, the boron neutrino flux is reduced by a *constant*, energy-independent factor.

Experiments will soon clarify the situation. The Gallium experiment, being performed at the Gran Sasso Laboratory and in the Soviet Union will distinguish possibilities I and II from III and IV. More Chlorine data can rule out the annual variation expected for just-so oscillations, or indicate a night-day effect such as explanation III might yield. Direct observations of solar neutrino scattering on electrons can provide a unique discrimination. The first results from Kamioka are expected in another year, and several more ambitious experiments are in the planning stage.

The evidence for the Fischbach *et al.* fifth force is scanty indeed: a questionable kaon decay experiment, the multiply-reanalyzed Eotvos experiment, and a two-sigma geophysical anomaly. Nonetheless, it is certainly possible that a new, relatively long-range, subgravitational force exists in Nature, even though nobody wants it. Many interesting new experiments have been inspired. Two objects of different composition may display different weights when compared both within and outside a deep mine. The Eotvos experiment can be repeated on an oblique terrain. The Cavendish measurement of Newton's constant can be repeated for various materials at high sensitivity. A quartz ball may fall at a slightly different rate than one of gold. The geophysical anomaly of Stacey *et al.* can be confirmed by measurements of gravity gradients under the sea. There are literally dozens of new experiments being proposed or performed, and the status of the fifth force will very soon be clarified.

A recent survey of the sky by Margaret Geller *et al.* seems to reveal an entirely unexpected result concerning the large-scale structure of the Universe. There are many large and roughly spherical voids in which there are no visible galaxies. Luminous galaxies congregate in the space between the voids, much as the soapy water forming the suds in the kitchen sink. The typical diameter of a Hubble-bubble is on the order of tens of megaparsecs. What weird mechanism has produced this curious distribution? Whatever it was, causality requires that it happened relatively recently in the history of the universe, probably after photon decoupling. Ostriker, Witten, Ruth Daly, and others have suggested various explosive origins to the Geller bubbles. Another possibility involves a low-energy phase transition. Perhaps the bubbles are regions of a different microphysical phase, or perhaps, they are relics of such regions. Possibly the phase transition has excluded baryons from the bubbles, just as salt is excluded when brine freezes. In another scenario, neutrinos are sufficiently massive in the voids so as to suppress nuclear fusion and accelerate stellar evolution. In this case, the voids are not at all void, but are populated with extinct stars. Despite the best efforts of my brilliant collaborators at Harvard, Boston University, and M.I.T., there doesn't seem to be any way to make sense out of this scenario. Not yet, anyway.

Suppose that the photon has a very small mass. This is an ugly though consistent modification of the standard (but not grand unified) theory. Now, suppose that an electrically charged particle falls into a black hole. What happens? Because the photon has a mass, electromagnetic gauge invariance is broken, and a black hole cannot acquire an electric charge. Thus, the flux of the fallen charge must vanish. It does so over a time interval given by the light travel time across the black hole. Thereby, charge conservation is violated. Now consider black holes that form in the centers of some galaxies. They are surrounded by a plasma of protons and electrons which they ingest. Normally, they would consume equal numbers of both particles in order to maintain electrical neutrality. If the photon has a mass, they are under no such compulsion. If they accrete more electrons than protons, the universe will slowly acquire a net positive charge. This effect could have significant cosmological consequences. This is the general subject of another endeavor, involving my student Eric Carlson and Aneesh Manohar. If it leads to anything interesting, I'll let you know next year. At least it's not strings.

CLOSING CEREMONY

The closing ceremony took place on Thursday, 14th August 1986. The Director of the School presented the Prizes and Scholarships as specified below.

PRIZES AND SCHOLARSHIPS

Prize for **Best Student**

awarded to Steve GIDDINGS, Princeton University, USA.

The Scholarships open for competition among the participants were awarded as follows:

Patrick M.S. Blackett Scholarship
Andrea PASQUINUCCI, University of Milano, Italy

James Chadwick Scholarship
Mark R. CARTER, Stanford University, USA.

Paul A.M. Dirac Scholarship
Giancarlo D'AMBROSIO, Harvard University, USA.

Amos De–Shalit Scholarship
Yigal SHAMIR, Tel Aviv University, Israel

Gunnar Källen Scholarship
Steven CARLIP, University of Texas, Austin, USA.

André Lagarrigue Scholarship
Paolo GIUBELLINO, CERN, Geneva, Switzerland.

Giulio Racah Scholarship
Michael McGUIGAN, Rockefeller University, USA.

J. John Sakurai Scholarship
John QUACKENBUSCH, University of California, Los Angeles, USA.

Giorgio Ghigo Scholarship
Alfred SHAPERE, University of California, Santa Barbara, USA.

Enrico Persico Scholarship

Peter FISHER, SIN, Villingen, Switzerland.

Peter Preiswerk Scholarship

Hillel LIVNE, Tel Aviv University, Israel.

Gianni Quareni Scholarship

Fernando QUEVEDO, University of Texas, USA.

Piersanti Matterella Scholarship

Gennaro MIELE, Istituto di Fisica Teorica, Napoli, Italy.

Olof Palme Scholarship

Zvi BERN, Lawrence Berkeley Laboratory, USA.

The following students received **Honorary Mentions** for their contributions to the activity of the School:

Krzysztof MEISSNER	Institute of Theoretical Physics, Warsaw, Poland
Peter VECSERNYES	Research Institute for Physics Budapest, Hungary

The following students received the **EPS** Scholarships:

Jan FTACNIK	Max-Planck-Institut, Munich, FRG
Roberto IASEVOLI	Dipartimento di Fisica, Napoli, Italy
Ivan KOSTOV	Institute for Nuclear Research, Sofia, Bulgaria
Krzysztof MEISSNER	Institute for Theoretical Physics, Warsaw, Poland
Peter NEUMAN	Institute for Theoretical Physics, Budapest, Hungary
Peter VECSERNYES	Central Research Institute, Budapest, Hungary

The following participants gave their collaboration in the scientific secretarial work:

Zvi BERN
Fabrizio BIANCHI
Jon BJORKMAN
Mark CARTER
Decio COCOLICCHIO
Giancarlo D'AMBROSIO
Peter FISHER
Steve GIDDINGS
Paolo GIUBELLINO
Stephen HAYWOOD
Ulriche KRAEMER
Juergen KRUEGER
Hillel LIVNE
Michael McGUIGAN
Krzysztof MEISSNER
Gennaro MIELE
Hans-Günter MOSER
Peter NEUMAN
Andrea PASQUINUCCI
John QUACKENBUSH
Fernando QUEVEDO
Thomas REDELBERG
Daniel ROHRLICH
Christopher SCOTT
Yigal SHAMIR
Alfred SHAPERE
Paul SHOTTON
Peter VECSERNYES
Stephen WEBB

PARTICIPANTS

Ahmed ALI	DESY Notkestrasse 85 2000 HAMBURG, FRG
Luis ALVAREZ-GAUMÈ	Physics Department Harvard University CAMBRIDGE, MA 02138, USA
Carlos AYALA	Department of Theoretical Physics Univ. Autonoma de Bellaterra BELLATERRA, BARCELONA, Spain
Luis BENTO	Centro de Fisica Nuclear Universidade de Lisboa Av. prof. Gama Pinto, 2 1699 LISBOA CEDEX, Portugal
Zvi BERN	Lawrence Berkeley Laboratory Theoretical Physics Group BERKELEY, CA 94720, USA
Stefano BERTOLINI	Department of Physics New York University NEW YORK, NY 10003, USA
Fabrizio BIANCHI	Istituto di Fisica dell'Università Via Massimo D'Azeglio, 46 10125 TORINO, Italy
Jon E. BJORKMAN	University of Colorado, Boulder Department of Physics High Energy Physics Group Campus Box 390 BOULDER, CO 80309, USA
Helmut BURKHARDT	CERN EP Division 1211 GENEVA 23, Switzerland

Tiziano CAMPORESI

CERN
EP Division
1211 GENEVA 23, Switzerland

Sergio CARACCIOLO

Scuola Normale Superiore
Piazza dei Cavalieri, 7
56100 PISA, Italy

Steven CARLIP

Department of Physics
The University of Texas at Austin
AUSTIN, TX 78712, USA

Mark R. CARTER

SLAC
P.O. Box 4349
STANFORD, CA 94305, USA

Decio COCOLICCHIO

Istituto di Fisica dell'Università
Via Amendola, 173
70126 BARI, Italy

Joanne COHN

The University of Chicago
The Enrico Fermi Institute
5640 S. Ellis Ave.
CHICAGO, IL 60637, USA

Francois CORRIVEAU

CERN
EP Division
1211 GENEVA 23, Switzerland

Guy COUGHLAN

Department of Theoretical Physics
University of Oxford
1 Keble Road
OXFORD, OX1 3NP, UK

Giancarlo D'AMBROSIO

Department of Physics
Harvard University
CAMBRIDGE, MA 02138, USA

David. J. DANIEL

Department of Physics
University of Edinburgh
King's Buildings
EDINBURGH EH9 EJZ, UK

Alessandro DE ANGELIS

Dip. di Fisica dell'Università
Via Marzolo, 8
35100 PADOVA, Italy

Lucia DI CIACCIO

Dipartimento di Fisica
Università Tor Vergata
Via Orazio Raimondo
00100 ROMA, Italy

Michael DINE

Physics Department
City College of New York
Convent Ave. at 138th St.
NEW YORK, NY 10031, USA

Cesareo A. DOMINGUEZ

ICTP
P.O. Box 586
34100 MIRAMARE - TRIESTE, Italy

Marcos DRACOS

Centre de Recherches Nucléaires
LEP-DELPHI
B.P. 20
67037 STRASBOURG CEDEX, France

Gosta EKSPONG

University of Stockholm
Institute of Physics
Vanadisvägen 9
11346 STOCKHOLM, Sweden

Jonathan EVANS

Department of Applied Mathematics
and Theoretical Physics
University of Cambridge
Silver Street
CAMBRIDGE CB3 9EW, UK

Pierre FAYET

Laboratoire de Physique Théorique
Ecole Normale Supérieure
24 Rue Lhomond
75231 PARIS, France

Sergio FERRARA

Physics Department
University of California
LOS ANGELES, CA 90024, USA

Ferruccio FERUGLIO

Dip. di Fisica dell'Università
Via Marzolo, 8
35100 PADOVA, Italy

Peter FISHER

SIN
5234 VILLIGEN, Switzerland

Jan FTACNIK

Max-Planck-Institut für
Physik und Astrophysik
P.O. Box 40 12 12
MUNICH, FRG

Reiner GARREIS

Institut für Theoretische Physik
Universität Karlsrhue
P.O. Box 63 80
KARLSRHUE, FRG

Steve GIDDINGS
Physics Department
Princeton University
P.O. Box 708
PRINCETON, NJ 08540, USA

George GINTHER
FERMILAB
P.O. Box 500
BATAVIA, IL 60510, USA

Paolo GIUBELLINO
CERN
EP Division
1211 GENEVA 23, Switzerland

Karl-Martin GLAS
Max-Planck-Institut für
Physik und Astrophysik
P.P. Box 40 12 12
MUNICH, FRG

Sheldon L. GLASHOW
Physics Department
Harvard University
CAMBRIDGE, MA 02138, USA

David GROSS
Physics Department
Princeton University
P.O. Box 708
PRINCETON, NY 08540, USA

Simon HANDS
Department of Physics
University of Edinburgh
EDINBURGH EH9 3JZ, UK

Stephen HAYWOOD
CERN
EP Division
1211 GENEVA 23, Switzerland

Timothy J. HOLLOWOOD
Department of Mathematical Sciences
University of Durham
South Road
DURHAM DH1 3LE, UK

Roberto IASEVOLI
Dipartimento di Fisica dell'Università
Mostra d'Oltremare - Pad. 20
80125 NAPOLI, Italy

Eric KAJFASZ
Centre de Physique des Particules
70, Route Léon Lachamp
13288 MARSEILLE CEDEX, France

Astri KLEGGE
University of Stockholm
Department of Physics
Vanadisvägen 9
11346 STOCKHOLM, Sweden

Ivan KOSTOV
Institute for Nuclear Research
and Nuclear Energy
Boulevard Lenin 72
1784 SOFIA, Bulgaria

Ulriche KRAEMMER
Institut für Hochenergiephysik
Akademie der Wissenschaften
Nikolsdorfergasse 18
1050 WIEN, Austria

Juergen KRUEGER
II Institut für Experimentalphysik
Notkestrasse 85
2000 HAMBURG 52, FRG

Francesco LACAVA
Istituto di Fisica
Università la Sapienza
Piazzale Aldo Moro, 5
00185 ROMA, Italy

Hillel LIVNE
School of Physics and Astronomy
Tel Aviv University
RAMAT-AVIV, TEL AVIV, Israel

Fedele LIZZI
HEP Division
Rutherford Appleton Laboratory
CHILTON, OXON OX11 OQX, UK

Ute LOEW
Institut of Theoretical Physik
University of Heidelberg
Philosophenweg 16
6900 HEIDELBERG, FRG

Dieter LUKE
DESY
Notkestrasse 85
2000 HAMBURG, FRG

Jnanadeva MAHARANA
CERN
TH Division
1211 GENEVA 23, Switzerland

Daniel MARLOW
Princeton University
Physics Department
P.O. Box 708
PRINCETON, NJ 08544, USA

Michael McGUIGAN
Physics Department
Rockefeller University
1230 York Avenue
NEW YORK, NY 10021, USA

Krzysztof MEISSNER — Institute of Theoretical Physics
Warsaw University
ul. Hoza 69
00–689 WARSAW, Poland

Robin MIDDLETON — Rutherford Appleton Laboratory
CHILTON, OXON OX11 OQX, UK

Gennaro MIELE — Istituto di Fisica Teorica dell'Università
Mostra d'Oltremare
80125 NAPOLI, Italy

Hans–Günter MOSER — CERN
EP Division
1211 GENEVA 23, Switzerland

Madhusree MUKERJEE — The University of Chicago
The Enrico Fermi Institute
5640 Ellis Avenue
CHICAGO, IL 60637, USA

Peter NEUMAN — Institute of Theoretical Physics
Eötvös University
Puskin Utca 5–7
10088 BUDAPEST VIII, Hungary

Keisuke OKANO — Universität Gesamthochschule Siegen
Fachbereich 7 – Physik
SIEGEN, FRG

René A. ONG — SLAC
P.O. Box 4349
STANFORD, CA 94305, USA

Ahmimed OURAOU — Département de Physique des
Particules Elémentaires
Centre d'Etudes Nucléaires de Saclay
91191 GIF–SUR–YVETTE CEDEX, France

Giampiero PAFFUTI — CERN
TH Division
1211 GENEVA 23, Switzerland

Michele PAOLUZZI — Dip. di Fisica dell'Università
Via Elce di Sotto
06100 PERUGIA, Italy

Andrea PASQUINUCCI — Dip. di Fisica dell'Università
Via Celoria, 16
20133 MILANO, Italy

John QUACKENBUSH	Department of Physics University of California LOS ANGELES, CA 90024, USA
Fernando QUEVEDO	Physics Department University of Texas AUSTIN, TX 78712, USA
Mariano QUIROS	CERN TH Division 1211 GENEVA 23, Switzerland
Stuart RABY	Los Alamos National Laboratory P.O. Box 1663 LOS ALAMOS, NM 87545, USA
Lars RASMUSSEN	CERN EP Division 1211 GENEVA 23, Switzerland
Thomas REDELBERGER	CERN EP Division 1211 GENEVA 23, Switzerland
Giovanni RIDOLFI	Istituto Nazionale di Fisica Nucleare Via Dodecaneso, 33 16146 GENOVA, Italy
Daniel ROHRLICH	Physics Department Brookhaven National Laboratory UPTON, NY 11973, USA
Carlo RUBBIA	CERN EP Division 1211 GENEVA 23, Switzerland
Philippe RUELLE	Institute de Physique Théorique Chemin du Cyclotron, 2 1348 LOUVAIN-LA-NUEVE, Belgium
Arthur SCHAFFER	CERN EP Division 1211 GENEVA 23, Switzerland
Julien SCHWINGER	Physics Department University of California LOS ANGELES, CA 90024, USA
Christopher SCOTT	Physics Department The University SOUTHAMPTON SO1 2PW, UK

Yigal SHAMIR	School of Physics and Astronomy Tel Aviv University RAMAT-AVIV, TEL AVIV, Israel
Alfred SHAPERE	Institute of Theoretical Physics University of California SANTA BARBARA, CA 93106, USA
Paul N. SHOTTON	CERN EP Division 1211 GENEVA 23, Switzerland
George SNOW	Dip. di Fisica dell'Università Via Irnerio, 46 40126, BOLOGNA, Italy
Luca STANCO	Dip. di Fisica dell'Università Via Marzolo, 8 35100 PADOVA, Italy
Urban STUDER	Institut für Hochenergiephysik ETH Hönggerberg 9093 ZURICH, Switzerland
Bo SUNDBORG	Institute of Theoretical Physics Chalmers Tekniska Högskola 412 96 GOTEBORG, Sweden
Joseph TARON	Departament de Fisica Teorica Universidad de Barcelona BARCELONA 08028, Spain
Peter VAN NIEUWENHUIZEN	Department of Physics State University of New York Stony Brook, NY 11794, USA
Peter VECSERNYES	Central Research Institute for Physics P.O. Box 49 1525 BUDAPEST, Hungary
Raul VIOLLIER	Institute for Theoretical Physics and Astrophysics University of Cape Town RONDEBOSCH 7700, South Africa
Stephen WEBB	The University Department of Theoretical Physics The Schuster Laboratory MANCHESTER M13 9PL, UK

Guy WORMSER
Université Paris Sud
Laboratoire de l'Accelerateur Linéaire
Centre d'Orsay
91405 ORSAY CEDEX, France

Shew-Ching YEH
CERN
EP Division
1211 GENEVA 23, Switzerland

Ming YU
Niels Bohr Institute
Blegdamsvej 17
2100 COPENHAGEN, Denmark

Peter ZIMAK
Institute for Theoretical Physics
and Astrophysics
University of Cape Town
RONDEBOSCH 7700, South Africa